Contemporary
Ceramic Formulas

Contemporary Ceramic Formulas

John W. Conrad, Ph.D.

Macmillan Publishing Co., Inc.
NEW YORK
Collier Macmillan Publishers
LONDON

Macmillan Publishing Co., Inc.
866 Third Avenue, New York, N.Y. 10022
Collier Macmillan Canada, Ltd.

Library of Congress Cataloging in Publication Data
Conrad, John W.
Contemporary ceramic formulas.
Bibliography: p.
1. Glazes—Formulae, tables, etc. I. Title.
TT922.C65   738.1'44   80–21926
ISBN 0–02–527640–9

10 9 8 7 6 5 4 3 2 1

Printed in the United States of America

This book is dedicated to my parents—W.L. (Bill) and Elizabeth Conrad.

Contents

Acknowledgments

The blending of the various clays, feldspars, fluxes, opacifiers, modifiers, and other ceramic minerals in innumerable combinations, for use at high and low temperatures, yields several thousand basic clay and glaze formulas. If one includes in the collection of formulas all the possible variations—however slight—in ceramic-material combination, the total number of formulas reaches well over 100,000. Before any formula can be recommended for use, of course, it must be tested to determine the acceptable qualities of surface, texture, fusion, plasticity, color, durability, transparency, strength, hardness, and other traits. For this book, more than one thousand formulas were tested, and those that produced the best results—and consistently so—were selected for presentation. Many, many hours of testing, retesting, analyzing results, and finally typing the manuscript were involved. For their help in this undertaking, special thanks go to Babs Baker, Anita Brashear, Linda Brunsardt, and Steven Patterson. The author is indebted, as well, to many other people for their advice, research, and interest in the writing of this book—colleagues, students, ceramists, and librarians, and to John Kent Baker and Jack Feltman. And to the staff at Macmillan is due grateful appreciation for their invaluable help in the organization, layout, and production of this book.

The photographs and drawings in the book were made by the author unless otherwise noted.

Introduction

When ceramists mold the raw, formless earth, they are doing more than just making pots or other wares. They are giving to the clay a "truth in structure," bringing forth the beauty that is born of a living relationship between humankind and nature. In past times, when each ceramic piece was made by hand, users could appreciate the kinship of potter and ware. In our own time, however, many people have been seduced by technology's ability to produce "perfect" pieces. In a poor bargain, the public may have unknowingly exchanged the ceramist's creativity for mass production's lifeless imitation. Today, it is the duty of the committed ceramist to remind the public that the potter's skilled hands can lend an imaginativeness to claymaking that an assembly line cannot.

For the serious ceramist, of course, extensive study, a tolerance for thoughtful planning and careful preparation, and simple hard work must lie behind any creative effort. It is the purpose of this book to provide the ceramist—the student, the dedicated amateur, the professional, the industrial potter—with both detailed, highly practical information on technique and a broad-ranging collection of ceramic formulas. The author believes that the more fully readers understand the ceramic media, and the wider the variety of formulas at their disposal, the greater the control and development of their ceramic craft. The material on procedure is intended to furnish readers with enough specific data on ceramic materials and equipment to enable them to experiment purposefully and knowledgeably, to develop their own formulas or adapt to their own needs and tastes those given in this book. It is the author's aim, in providing information on the construction of equipment, to present readers with an alternative to what may be expensive and/or difficult-to-obtain commercially produced goods.

Although some of the topics covered in this book are complicated and highly specialized, other areas, such as clay bodies, refractory equipment, and raw stains, can be readily understood even by the beginning student. The author has assumed that the reader knows how to measure, mix, and use ceramic formulas and has some basic knowledge of the ceramic process. Readers who do not have such expertise or who need to brush up on their skills are referred to *Ceramics,* by G. Nelson; *Clay and Glazes for the Potter,* by D. Rhodes; or *Ceramic Manual,* by J. Conrad (see bibliography). It is recommended, too, that the inexperienced student take a course in clay/ glaze technology, obtain the assistance of an experienced ceramist, and/or read extensively on clays and glazes. For the beginner to select a glaze formula at random, mix it, slap it on a pot and fire it may lead to disaster—a spoiled pot, a broken kiln shelf, a battered or ruined kiln. To the experienced potter the author emphasizes the need for constant, well-planned, and well-carried-out testing at all stages in the ceramic process. Testing is essential because each ceramist has his or her own methods of weighing, blending, and using ceramic formulas, and because requirements and preferences vary widely, as does equipment available.

This book is devoted to ceramic technique rather than to aesthetics; nevertheless, the author believes that a skilled hand must have a discriminating eye to guide it. Forms that are chiefly functional can be visually marred by details that, however practical, do not complement the ware as a whole. Conversely, forms that are primarily sculptural but that represent little more than sheer surface pyrotechnics or the emotional gratification of the potter may not achieve the true goal of the ceramic art: the integration of form and surface decoration. Today it is becoming fashionable to accept anything that falls off the wheel—whether fired or not—and to present it as a creative endeavor. Yet many such pieces may well be unsatisfactory in themselves, or inappropriate to the purpose they were intended to fulfill.

Contemporary
Ceramic Formulas

I

Clay

Earthenware, Mid-Temperature Bodies, Talc Body, Stoneware, Ovenware, Low-Shrinkage Bodies, Self-Glazing Bodies, Porcelain

The term *clay* in ceramics refers to a stiff viscous earth material of many varieties, found in beds or deposits near the surface of the earth and various depths below; or to synthetic blends of various mineral compositions. Clay, when added to water, forms a tenacious paste capable of being molded or thrown on a wheel. With additional water, it can be made into a *slip,* to be cast in a porous mold. The form hardens when dry and, when subjected to high heat, is converted into a rock-like material. *Pottery, stoneware, earthenware, porcelain, whiteware, raku,* and *ironstone* are categories of claywork based on the temperature at which the clay matures and on the composition of the clay from which an object is made. Objects made from clay include dinnerware, teapots, vases, bricks, mugs, goblets, serving pieces, sewer pipe, tile, and the like.

The Formation of Clay

The elements that make up clays are an important aspect of the formation of the earth's surface. Clays were and still are formed under intensive conditions which involve, at one extreme, the action of compressed water vapor at several hundred degrees Fahrenheit, and, at the other extreme, the action of atmospheric agencies and mineral movement at ordinary temperatures. Clays were not part of the original magmatic development of molten matter, which made up the earth's initial crust and became igneous rock after cooling. There is, however, a direct relationship between magmatic development and clay formation. During magmatic formation, the fluids and vapors of the magma interacted with the igneous rock to form clay mineral masses. This process is the extreme level of development of high-temperature clay mineral formation. Low-temperature transition, during which the acids, normal temperatures, and pressure alter the minerals, is gradual. In time, though, the clay aggregates of hydrothermal origin become difficult to distinguish from clay aggregates formed under atmospheric conditions created by the movement of minerals by water, air, and volcanic eruptions.

The clay minerals formed by *hypogene* process (occurring below the earth's surface) are the result of the actions of gases, vapors, or solutions that force their way upward through the rocks of the earth's crust. The primary minerals removed from the crustal rocks by the action of gases, vapors, solutions, and water are alumina, silica, alkali, and iron. They are transformed into clay minerals at temperatures ranging from 140°F to 750°F in various acid, neutral, or alkaline environments, depending upon the pH of the invading rocks, vapors, and solutions.

The process by which clays are formed on or near the earth's surface is the result of atmospheric conditions that occur under a variety of circumstances. Some clay minerals developed under normal surface conditions during geologic periods that ranged from short to long duration. The production of other clay minerals, however, was accelerated by the concentration of favorable conditions. The leaching, depositing, and weathering (the effects of water and/or air movement in the environment) of the materials—called the *supergene* process—involves complex chemistry and occurs very slowly even under ideal conditions. Even weak concentrations of acid, or of alkali together with alkali or alkaline-earth elements (alumina, silica, etc.), when given sufficient time, will produce clay deposits, some of which are very large. Special conditions, such as more concentrated chemical action (alkali or acid) and/or a moderate increase in temperature, will greatly reduce the time factor. In brackish or saline bodies of water, the clay minerals are subject to *diagenesis* (transformation by the dissolution and recombination of the elements). In regions of heavy rainfall and high temperatures, organic acids are apt to accumulate in greater-than-normal concentrations. In bogs, where the temperature is not extreme and the concentration of organic acids is not high, clay minerals will be produced, although at a much slower rate.

The formation of clay minerals, by either hypogene or supergene methods, followed by their removal, transportation, and redeposition, is a worldwide occurrence. Clays can be broadly characterized as residual and sedimentary. *Residual clays* are those, such as china clay, that remain in the geographical region in which they were formed. They are found in stratiform deposits that lie on or near the earth's surface. *Sedimentary clays,* or clays that have been transported from their place of origin, accumulate on floors of lakes and oceans, in certain glacial deposits, in desert basins, in river deltas, and in windblown deposits. An example of sedimentary clay is ball clay.

The Makeup of Clay

On the basis of its mineral constituents, clay can be classified into four groups: kaolin, montmorillonite, hydromica, and palygorskite. The various groups seldom occur in a pure state; usually the constituent materials are mixed with impurities of quartz, feldspar, iron, and/or organic matter. *Kaolin* group contains several clay minerals that have very similar chemical compositions ($Al_2O_3 \cdot 2\ SiO_2 \cdot 2\ H_2O$) that form six-sided plate-like crystals. *Montmorillonite* group $2[(Al_{1.67}Mg_{0.33})Si_4O_{10}(OH)_2]$ often contains sodium calcium varieties and traces of zinc or iron. Montmorillonite clay has a flake-like structure and forms collidal gel (thixotraphic). *Hydromica* group is the most complex of the clays and has a layer-lattice structure; hydrous micas [$Al_2K(Si_{1.5}Al_{0.5}O_5)_2(OH)_2$] are the most commonly used clay in this group. The principal mineral of *palygorskite* group is attapulgite. Sepiolite, the best-known clay in this group, is used in the manufacture of meerschaum smoking pipes and filters. The minerals in these clays have a chain-like atomic structure.

There is a great deal of variation, in contemporary ceramics, in types of clay bodies used. Very few clays, as they come out of the ground, have all the desired properties for any given purpose. Rather, various clays, alone or in mixtures, are selected for their particular contribution to the clay body: handling properties, drying strength, dry strength, and fired properties. In some cases clays are added to the clay body mix because they offer an inexpensive body constituent (filler), in others because they contain or will produce a desired chemical composition or a special physical attribute. Most clay bodies used for throwing, hand building, or casting, for example, are a blend of several clays, feldspars, colorants, and silicas. Moreover, each type of clay body is highly individualistic and will vary, however slightly, from shipment to shipment. The wide range of shaping methods creates a further need for variation in the makeup of clay bodies. The more than half-dozen shaping methods used in both studio and factory include dust press (tile), stiff-mud (brick), hot pressing (bearings, machinery parts), soft-mud molding (jiggering), throwing (potter's wheel), hand building (sculpture), and slipcasting (dinnerware). In addition, individuals, studios, and industry use clays in varying amounts and in different conditions of flocculation (the amount of water in the clay). The same clay that would be considered satisfactory for one process might not be suitable for another. Finally, two different studios or potters, when evaluating the properties of a particular clay body, are likely to use differing criteria to determine whether the material meets their physical and/or aesthetic requirements. No wonder there is such a variety of clay bodies in use today.

Each clay has a unique property that contributes to the clay body. Table 1 lists the various properties of several common clays—plasticity, softening point, shrinkage, absorption. Some clays, such as brick clay and natural stoneware, can be used straight from the ground. These clays then require only the minimal processes of crushing and screening. The moist clays used by potters for throwing and hand building are, on the other hand, blends of clays that, when

Table 1
Properties of Some Common Clay Types

Type	Plasticity	Softening Point	Origin	Fired Color	Absorption	Shrinkage	Grain	Use
Kaolin Avery Kingsley Monarch	Low	1650°C	Mainly residual	Strong white	Low	Low	Fine	Casting porcelain, white clay bodies
E.P.K.	Medium	1650°C	Sometimes sedimentary	White	Low	Medium	Fine	Casting white throwing bodies
Ball Clay Bandy C & C Imperial Kentucky Rex Tennessee	High	1225° to 1425°C	Sedimentary	Off-white to medium gray	Moderate	High	Fine	Very plastic, throwing bodies
Fire Clay A. P. Green Monarch North American	Low	1225°C	Sedimentary	Tan to orange-brown	Low	Moderate	Coarse	Kiln shelves (mullite)

Table 1 (continued)

Type	Plasticity	Softening Point	Origin	Fired Color	Absorption	Shrinkage	Grain	Use
Stoneware (natural) Goldart Jordon Ohio Sagger	Medium to high	1280° to 1300°C	Sedimentary	Light tan, gray, or light brown	Low	Medium to high at C/9	Medium	Dinnerware, small sculpture
Brick Shale brick	Low	1090°C average	Sedimentary	Red to brown	Low	Moderate	Whole range	Bricks, tiles, flowerpots, roofing tiles, sewer pipes
Earthenware Cedar Heights	Medium	1180°C	Sedimentary	Red to brown	High	Moderate	Fine to medium	Terra-cotta, pottery, low-temperature figurines, colorant for stoneware

mixed with sufficient water, will form a mass that exhibits typical plastic flow having yield stress, and a high degree of extendibility (the amount of deformation the clay can take before it begins to rupture).

Modifications of the working properties of clay have been traced to the earliest potters, who felt the urge to experiment with clay because it was often impractical or uneconomical to discard an unsatisfactory clay and search for a new or better one. Archeological evidence shows, for example, that in some ancient ware, pulverized rock was added to excessively sticky clay. The present-day ceramist would do well to become familiar with the various ingredients that make up clay bodies and with the properties that each ingredient contributes to the body. Table 2, which outlines some commonly used ingredients and their properties, can serve as a guide to the ceramist in formulating a new clay body or in modifying an existing one.

A knowledge of clay types and their properties should enable the ceramist, furthermore, to detect some of the most frequently occurring problems and to correct them easily, since defects can ordinarily be eliminated by alteration of the body formula. There are, in fact, three basic approaches to the modification of body formulas:

1. *Adding materials,* such as 5 to 10 percent grog and 5 to 20 percent fire clay (to stiffen the clay and reduce shrinkage and warpage).
2. *Reducing the amount of minerals* in the body by using 10 percent ball clay, half of the bentonite, 20 percent of the earthenware, or 5 to 15 percent of the silica and flux.
3. *Substituting materials,* as, for example, replacing half of the plastic kaolin or fire clay with earthenware or ball clay, using grog instead of silica sand, and/or replacing a third of the silica with kaolin.

All, part, or a combination of these methods will improve the clay body. Table 3 illustrates the point in chart form.

Testing Clay Bodies

Hundreds of clay body formulas are available. Each formula should be tested to account for the inclusion of local materials, to check

Table 2

Clay and Materials and Their Use in Clay Bodies

Material	Use	Earthenware	Percent Stoneware	Porcelain
Kaolin	White coloring agent, high-temperature resistance	10 to 50	0 to 30	10 to 50
Ball Clay	Plasticizer	0 to 30	0 to 30	0 to 30
Fire Clay	Filler, coloring agent, graining agent, stiffener	0 to 20	0 to 35	—
Earthenware Clay	Coloring agent, filler	0 to 80	0 to 40	—
Bentonite	Plasticizer	0 to 5	0 to 5	0 to 5
Colorants (iron, ilmenite)	Coloring agent, texturizer	0 to 10	0 to 10	—
Flux (feldspars)	Vitrifier	0 to 30	0 to 20	10 to 30
Flint	Hardener, stiffener	0 to 25	0 to 20	20 to 25
Grog or Sand	Stiffener, body opener	0 to 10	0 to 15	0 to 5

Table 3
Clay Body Problems and Their Correction

Problem	Correction
Too sticky	Decrease ball clay or add fire clay
Too gritty	Screen, or use less gritty clay or less sand and/or grog
Not plastic enough	Add ball clay and/or bentonite
High shrinkage and/or warping	Reduce ball clay and/or earthenware; add fire clay
Fired clay brittle	Fire at lower temperature, increase kaolin and silica, or decrease flux
Low vitrification temperature	Add kaolin and/or silica
Color too dark	Reduce colorants, change fire clay, increase light-color-bodied clays
Color too light	Add or increase colorants

the firing techniques, to ascertain what the clay body properties are, and to determine how the glaze will respond to the body. Testing 100 grams of a clay body can be done quickly and easily. If the formula being used is a batch formula (in which the total amount of the materials may add up to more than or less than 100), then it is necessary to convert the measurements to percent by weight. The conversion will bring all the materials in the formula to between 99.9 and 100.1 percent by weight. The percent-by-weight method makes calculations of formula alterations and/or colorant changes easier.

Remember, the materials in the base formula are calculated so that the total is 100 grams plus or minus 0.1. The weighed materials are placed in a 6-to-8-ounce foam or paper cup, and a sufficient amount of water is added to produce a throwing-on-the-wheel consistency. After soaking, the mixture is removed from the cup and blended by hand on the wedging board. The first test to be carried out is for plasticity. The plasticity of clay is determined by particle shape and size, type of clay mineral, and the relative presence in the clay of soluble salts, absorbed ions, and organic matter. If the test clay is too dry, additional water is added; if too wet, additional wedging is done. After wedging, half of the clay is formed into a rope, whose diameter is continually reduced by rolling on a table. Occasionally the rope is bent like a horseshoe to determine if it will crack. When the rope cracks, the diameter of the clay is measured and recorded. The smaller the diameter of the clay, the more plastic the clay is considered to be. As a reference, the plasticity of the clay bodies listed in this book has been rated. A rating of 2 to 6 is adequate for hand building; of 2 or better is sufficient for casting; and of 6 or better is sufficient for throwing. The scale used as a guide for judging plasticity follows:

Table 4
Clay Plasticity Scale

Number	Plastic Quality
0	Not plastic, very hard to use; 2″ rope breaks easily
2	Dry, crumbles easily; 1″ rope breaks easily
4	Short, cracks easily; ¾″ rope will bend
6	Adequate for hand building; ½″ rope will bend
8	Plastic, good for throwing; ⅜″ rope will bend
9	¼″ rope will bend
10	Very plastic; ¹⁄₁₆″ rope will bend

To carry out the next test, for shrinkage, the clay is blended and formed into a test bar shape (Drawing 1) that is about 5 inches long, 1½ inches wide, and ⅜ inch thick. Two marks, 10 centimeters apart, are made in the moist clay bar, and a hole is put in at one end. The bar is set out to dry; after drying, it is bisque-fired. A favorite glaze is then applied across the middle of the bar; then the bar is placed in the kiln and fired to recommended temperature. When the bar is cool, the two marks are measured to determine the percentage of shrinkage. In measuring the fired clay, the original notches placed in the moist clay should be considered to mark off 100 units (that is, 100 millimeters). To calculate the amount of shrinkage, the "0" end of a metric ruler should be placed at one mark; then the number of units between the second mark and the 10-centi-

Drawing 1
Top and Cross-Section Views of Clay Test Bar

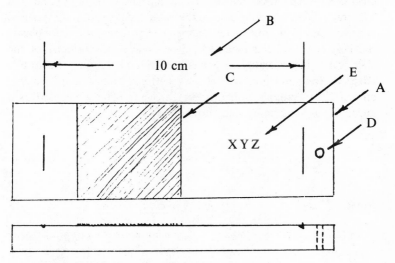

A. Test bar (100 grams, dry weight)
B. Two marks in the moist clay, 10 centimeters apart
C. Glaze painted on the clay surface
D. Hole through the clay for stringing together or for display
E. Identification key

meter mark on the ruler should be counted. The figure arrived at will represent the percent of shrinkage. For example, if the number of units between the second mark and the 10-centimeter mark is 14, then the percent of shrinkage is also 14. The test for shrinkage is now completed.

In preparation for the following test on absorption, or vitrification, the dry bar is weighed and then placed in a container of water for twenty-four hours. It is then patted dry and reweighed. The percent of absorption can be determined from the difference between the two weights.

The simple test sequence outlined will provide the ceramist with adequate information to decide whether the clay body has sufficiently favorable properties to warrant the weighing and preparation of large quantities of it. The major qualities determined from this clay testing include:

1. degree of plasticity
2. grittiness or sandiness of the clay body texture
3. percent of shrinkage
4. absorption (vitrification)
5. ability of the glaze to fuse to the clay
6. fired color
7. maturation temperature
8. surface quality and texture
9. effects of the clay on the glaze

Formula Filing

With over 8,000 ceramic formulas in his collection, covering all aspects of ceramics, the author has devised a method of organizing, categorizing, and recording information; of avoiding duplication; and of filing, all by using the *keysort card system.* Purchased cards come prepunched with holes near the edges. By cutting the edge of the card through to a particular hole or holes, the ceramist indicates the chief data the card contains. Then, when a specific card or piece of information is desired, a long needle is inserted through the appropriate hole or set of holes, the cards are vibrated, and the desired card or cards drop from the stack. A master list (Table 5) containing all the data—on temperatures, sources of information, types of formulas, and other related details—is used to select the correct holes to be cut through in new cards. Table 5 shows a stoneware-glaze formula and pertinent data on a sample keysort card, as well as the master list itself.

It is recommended that the reader, in picking out a formula, consider—in addition to such traditional factors as temperature range and qualities desired in the clay body—the cost, reliability, and availability of materials. Local and readily accessible materials are often cheaper and more dependable than materials shipped long distances. Further, some ceramic materials are not always available locally, for several reasons: shipping costs are high; local materials are more popular than imported ones; some mining companies have stopped producing certain ceramic materials; and local ceramic supply houses

Table 5

Keysort Card, Ceramic Data, and Example of Glaze and Its Information

Master List of Data

Clay
Engobe and slip
Colorants
Stain

Melted glass
Glass
Decal
Silkscreen

Glaze
Kiln wash
Fuming
Crystal

Underglaze
Slip glaze
Egyptian paste
Adhesive

Kiln wadding
Used with salting
Low shrinkage
Salt mix

Enamel
Frit
Leadless
Substitute

Brick, castable
Earthenware C/022-04
Mid-temperature C/03-3
Stoneware C/4-10

Porcelain C/6-up
Raku
Salt
Ovenware

Self-glazing clay

Source
Book II
Manual
Contemp. Ceramics
Book I
Books or magazines
Ceramists

R12 R11 R10 R9 R8 R7 R6 R5 R4 R3 R2 R1

Jack Pott

G3066 El. fire "Sahara"

Leadless Glaze

Temperature	C/8
Surface @ C/8	Semigloss
Fluidity	None
Stain penetration	None
Opacity	Opaque
Color/oxidation	Dark eggshell

Potash feldspar	24.6
Whiting	22.6
Flint	18.5
Kaolin	9.2
Rutile	8.1
Nepheline syenite	5.8
Zinc oxide	3.3
Zircopax	3.3
Red iron oxide	2.8
Bentonite	1.7
	99.9

McBEE ● SYSTEMS ● ATHENS, OHIO

KSS 3711N 503

do not stock an extensive inventory of materials. And, regardless of materials used, ceramists should be sure to test all new formulas.

Earthenware

Earthenware is an English term for pottery that is not vitrified. Many different clay bodies are classified as earthenware; even many stoneware or porcelain clays, when fired at earthenware temperatures, can be so classified.* In general, earthenware clay is porous, nonwaterproof (unglazed), opaque, unvitrified, medium-strong, and fired in the C/010 to C/02 temperature range. Earthenware is grouped according to type of glaze, decoration, composition, and maturation temperature.

In different cultures, time periods, and worldwide locations, many types of earthenware have been produced. The following are the best known: *Creamware* is a cream-colored earthenware with a transparent glaze, developed by Josiah Wedgwood (1760) to compete with porcelain; *Delftware* refers to a tin-glazed earthenware made in Holland and England during the seventeenth and eighteenth centuries. An earthenware in which the entire surface is burnished with a polishing stone while the clay is in the leather-hard stage is called *Burnishedware*. This method of smoothing the clay surface is common to cultures all over the world, from the most primitive village potters to the most advanced industrial manufacturers, from antiquity to the present time. *Faience* is a term used for a tin-glazed earthenware, especially that made in France, Germany, and Scandinavia. This French term is derived from the name of the Italian town Faenza, where the earthenware is made. The faience technique is the same as that of delftware and *maiolica*, the only difference being place of origin; *maiolica* is made in Italy. (*Majolica* is the anglicized version of the Italian term but it more correctly refers to an earthenware, glazed with semitransparent glazes, made in England during the nineteenth century.) *Pearlware* is a type of earthenware that Wedgwood produced (1779) by including a greater percentage of silica and white clay to the body, making it effective for underglazing

---

* It is desirable for stoneware or porcelain clays to be improved by the addition of a flux (colemanite, frit, soda feldspar, etc.) to lower the vitrification point.

C96* Sculptural Stoneware

Fire clay	40^{λ}		Temperature	C/6–9	(The maturing temperature range for the clay)
Ball clay	30		Shrinkage	8%	(The percentage of clay shrinkage from moist to fired condition)
Grog, 20 mesh	12		Glaze scale	9	(The indication of the effects the clay has on a translucent glaze)
Flint	10		Plastic scale	4	(The indication of the clay plastic qualities as stated in the scale in Table 4, page 10)
Grog	8		Color / oxidation / reduction	Tan brown Brown	(The clay color as a result of firing in a reduction or oxidation firing atmosphere)
Manganese dioxide	3				
	103				

* The clay body number used for organization and identification.
λ The formulas listed are percentage by weight.

coloring and giving brighter colors. *Pre-Columbian* ware was made in the Americas before Columbus, using hand-turned wheels, press molds, and coiling. These containers were decorated with slips, polished, and then fired in both reduction and oxidation. *Raku,* originally a Korean development, was later expanded by the Japanese. This soft and porous ware is covered with a lead glaze, fired at a low temperature, and then removed from the kiln at peak temperature to air-cool quickly. *Terra-cotta* is a red earthenware with a highly grogged body that is often unglazed and used for sculpture.

An example of a stoneware clay formula that includes an explanation of related data which will help the reader understand subsequent formulas is on page 15.

The following formulas are for various usages and colors. The high-grog-content clays are for hand building, for sculpture, or for thick-walled ceramics. Colors include white, tan, buff, brick red, orange, pink, and gray.

EARTHENWARE CLAY BODY FORMULAS

CB 1001 Burnt Orange Brick

Local red clay	55	Temperature	C/010*
Frit #25 PEMCO	20	Shrinkage @ C/010	13.0%
Silica	15	Plastic scale	8.7
Talc	10	Color	Burnt orange brick (80**)
	100	Absorption	9.8%

CB 1002 Pinkish-Tan Earthenware

Cedar Heights goldart	65	Temperature	C/08
Fire clay	20	Shrinkage @ C/08	7.0%
Ball clay	10	Plastic scale	8.0
Frit #3110 FERRO	5	Color	Light pinkish-tan (40)
	100	Absorption	1.5%

* The temperature scale used in ceramics for the pyrometric cones covers a range from the lowest temperatures—C/022, 021, 020, 019—to the highest—C/01, 1, 2, 3, 4, to 11, 12, 13, etc.

** Numbers in parenthesis refer to swatches in the color insert following page 80.

CB 1003 Pinkish-White Earthenware

Tennessee ball	60	Temperature	C/08–04
Nepheline syenite	20	Shrinkage @ C/08	9.0%
Fire clay	10	Plastic scale	8.5
Talc	10	Color	Pinkish-white (63)
	100	Absorption	15%

CB 1004 Oaktag Earthenware

Ball clay	50	Temperature	C/08
Talc	25	Shrinkage @ C/08	12.0%
Frit #3110 FERRO	15	Plastic scale	1.0
Silica sand, fine	10	Color	Oaktag (64)
	100	Absorption	8.3%

CB 1005 Light Tan Earthenware I

Plastic fire clay	50	Temperature	C/08–04
Fine grog	15	Shrinkage @ C/08	9.0%
Medium grog	15	Plastic scale	8.0
Gerstley borate	10	Color	Light tan (62)
Ball clay	10	Absorption	14.7%
	100		

CB 1006 Light Tan Earthenware II

Fine grog	40	Temperature	C/08–04
Ball clay	30	Shrinkage @ C/08	9.6%
Plastic fire clay	20	Plastic scale	8.0
Silica sand	6	Color	Light tan (40)
Bentonite	4	Absorption	16%
	100		

CB 1007 Pinkish-White Earthenware

E.P. kaolin	40	Temperature	C/08–06
Talc	20	Shrinkage @ C/08	9.0%
Nepheline syenite	20	Plastic scale	6.0
Ball clay	20	Color	Pinkish-white (63)
	100	Absorption	9.7%

CB 1008 Brick Orange Earthenware I

Cedar Heights redart	100.0	Temperature	C/06–04
	100.0	Shrinkage @ C/06	Not measured
Soda ash	0.4	Plastic scale	9.5
Sodium silicate	0.2	Color	Brick orange (80)
		Absorption	7.9%

CB 1009 Red Earthenware

Sewer pipe clay	90	Temperature	C/06–04
Fine grog	10	Shrinkage @ C/06	9.0%
	100	Plastic scale	2.0
		Color	Deep brick orange (80)
		Absorption	12.7%

CB 1010 Brick Orange Earthenware II

Sewer pipe or		Temperature	C/06–04
brick clay	70	Shrinkage @ C/06	8.0%
Grog	10	Plastic scale	1.5
Ball clay	10	Color	Brick orange (80)
Talc	5	Absorption	13.3%
Nepheline syenite	5		
	100		

CB 1011 Pinkish-Tan Earthenware II

Kentucky old mine #4	60	Temperature	C/06–02
Volcanic ash	25	Shrinkage @ C/06	8.0%
Plastic fire clay	15	Plastic scale	9.0
	100	Color	Light pinkish-tan (40)
		Absorption	16.1%

CB 1012 Coral Earthenware

Cedar Heights bonding		Temperature	C/06–04
clay (50 mesh)	52	Shrinkage @ C/06	6.0%
Medium grog	22	Plastic scale	4.0
Cedar Heights redart	13	Color	Coral (59)
Tennessee ball	13	Absorption	13.3%
	100		

CB 1013 Light Pink Earthenware

Fire clay	50	Temperature	C/06–02
Ball clay	50	Shrinkage @ C/06	12.0%
	100	Plastic scale	9.3
		Color	Light pink (44)
		Absorption	17.3%

CB 1014 Navajo White Earthenware

Nepheline syenite	50.0	Temperature	C/06–02
E.P. kaolin	20.0	Shrinkage @ C/06	16.0%
Kentucky ball #4	17.0	Plastic scale	8.0
Flint	13.0	Color	Navajo white (30)
Sodium silicate	0.2	Absorption	13.9%
	100.2		

CB 1015 Oaktag Earthenware II

Kentucky ball	45.0	Temperature	C/06–02
Talc	30.0	Shrinkage @ C/06	13.0%
E.P. kaolin	15.0	Plastic scale	6.5
Silica	10.0	Color	Oaktag (31)
	100.0	Absorption	15.5%
Sodium silicate, dry	0.3		

CB 1016 Light Coral Earthenware

Fire clay	45	Temperature	C/06–04
Kentucky ball #4	15	Shrinkage @ C/06	14.0%
Talc	15	Plastic scale	8.5
Silica sand	10	Color	Light coral (59)
Medium grog	10	Absorption	12.1%
Colemanite	5		
	100		

CB 1017 Pink Earthenware

Greenstripe fire clay	40	Temperature	C/06–02
Lincoln fire clay	20	Shrinkage @ C/06	10.0%
Kentucky ball	10	Plastic scale	8.3
Silica sand	10	Color	Light pink (44)
Sawdust ⅛″ mesh	10	Absorption	17.8%
Nepheline syenite	8		
Bentonite	2		
	100		

CB 1018 Casting Earthenware

E.P. kaolin	40	Temperature	C/06–04
Kentucky ball #4	25	Shrinkage @ C/06	10.0%
Frit #25 PEMCO	10	Plastic scale	2.0
Flint	10	Color	Navajo white (6)
Talc	10	Absorption	12.5%
Fine grog	5		
	100		

CB 1019 Jordon Earthenware

Jordon clay	30	Temperature	C/06–04
Redart	30	Shrinkage @ C/06	8.0%
Plastic fire clay	27	Plastic scale	9.5
Fine grog	10	Color	Brick orange (68)
Red iron oxide	3	Absorption	12.8%
	100		

CB 1020 Monmouth Clay Earthenware

Mixed grog	30	Temperature	C/06–04
Monmouth clay	25	Shrinkage @ C/06	8.0%
Red clay	20	Plastic scale	8.5
Fire clay	19	Color	Deep brick orange (80)
Bentonite	2	Absorption	13.9%
Red iron oxide	4		
	100		

CB 1021 Medium Coral Earthenware

Local brick or		Temperature	C/06–04
sewer pipe clay	30	Shrinkage @ C/06	14.0%
Local stoneware clay	30	Plastic scale	8.5
Silica sand	20	Color	Medium coral (59)
Kentucky ball #4	10	Absorption	12.6%
Soda feldspar	10		
	100		

CB 1022 Plastic Earthenware

Plastic fire clay	25	Temperature	C/06–02
Ball clay	25	Shrinkage @ C/06	12.0%
Local red earthenware	25	Plastic scale	9.0
Fine grog	10	Color	Brick orange (68)
Medium grog	10	Absorption	14.4%
Gerstley borate	5		
	100		

CB 1023 Light Orange Earthenware

Plastic fire clay	24	Temperature	C/06–02
Jordon clay	24	Shrinkage @ C/06	14.0%
Silica sand	20	Plastic scale	9.3
Fine grog	20	Color	Light orange (61)
Talc	10	Absorption	13%
Bentonite	2		
	100		

CB 1024 Off-White Earthenware

Kentucky ball #4	58	Temperature	C/04–02
China clay	30	Shrinkage @ C/04	11.0%
Lead monosilicate	12	Plastic scale	9.3
	100	Color	Off-white (31)
		Absorption	13.7%

CB 1025 Low-Absorption Earthenware

Ball clay	55	Temperature	C/04
Frit #25 leadless	15	Shrinkage @ C/04	12.0%
Talc	15	Plastic scale	8.0
Nepheline syenite	15	Color	Oaktag (31)
	100	Absorption	5%

CB 1026 Jordon Clay Earthenware

Jordon clay	55	Temperature	C/04–2
Red clay	20	Shrinkage @ C/04	12.0%
Plastic fire clay	15	Plastic scale	9.5
Black barnard clay	10	Color	Deep brick orange (80)
	100	Absorption	14.5%

CB 1027 Earthenware

Ball clay	55	Temperature	C/04–02
Jordon clay	10	Shrinkage @ C/04	12.0%
Nepheline syenite	10	Plastic scale	9.5
Talc	10	Color	Light pinkish-tan (40)
Silica	15	Absorption	11.4%
	100		

CB 1028 Gray-Blue Earthenware

Ball clay	55	Temperature	C/04
Frit #25 PEMCO	20	Shrinkage @ C/04	13.0%
Talc	9	Plastic scale	9.5
Plastic fire clay	8	Color	Light gray-blue (7)
Red iron oxide	3	Absorption	0.4%
Cobalt oxide	2		
Manganese dioxide	2		
Copper black oxide	1		
	100		

CB 1029 Brick Brown Earthenware

Jordon clay	50	Temperature	C/03–02
Redart	15	Shrinkage @ C/03	11.0%
Fire clay	15	Plastic scale	9.0
Barnard clay	10	Color	Brick brown (80)
Fine grog	5	Absorption	11.6%
Talc	5		
	100		

CB 1030 Ochre Earthenware I

Plastic fire clay	50	Temperature	C/04–02
Spodumene	15	Shrinkage @ C/04	10.0%
Fine grog	15	Plastic scale	9.0
Ball clay	20	Color	Light ochre (62)
	100	Absorption	11.5%

CB 1031 Ochre Earthenware II

Fire clay	50	Temperature	C/04–2
Spodumene	30	Shrinkage @ C/04	5.0%
Talc	10	Plastic scale	2.0
Fine grog	10	Color	Light ochre (62)
	100	Absorption	17.1%

CB 1032 Eggshell White Earthenware

Ball clay	43.8	Temperature	C/03–2
China clay	28.0	Shrinkage @ C/03	8.0%
Flint	19.8	Plastic scale	9.3
Cornwall stone	8.4	Color	Eggshell white (6)
	100.0	Absorption	18.4%

CB 1033 Earthenware I

Ball clay	42	Temperature	C/04–2
China clay	30	Shrinkage @ C/04	6.0%
Flint	19	Plastic scale	9.0
Medium grog	9	Color	Off-white (30)
	100	Absorption	17.8%

CB 1034 Earthenware II

Local red clay	30	Temperature	C/04–2
Ball clay	30	Shrinkage @ C/04	9.0%
E.P. kaolin	20	Plastic scale	9.5
Fine grog	10	Color	Medium coral (59)
Flint	10	Absorption	18.4%
	100		

CB 1035 Earthenware III

Flint	30	Temperature	C/04–2
E.P. kaolin	25	Shrinkage @ C/04	10.0%
Ball clay	20	Plastic scale	6.0
Custer feldspar	20	Color	Brick (80)
Red iron oxide	5	Absorption	18.8%
	100		

CB 1036 Pinkish-White Earthenware

Ball clay	30	Temperature	C/04
Whiting	30	Shrinkage @ C/04	7.0%
China clay	20	Plastic scale	8.5
Fine white sand	20	Color	Pinkish-white (63)
	100	Absorption	13.9%

CB 1037 Dense Earthenware

Plastic fire clay	68	Temperature	C/02
Monmouth clay	15	Shrinkage @ C/02	Not measured
Soda feldspar	17	Plastic scale	9.3
	100	Color	Light ochre (62)
		Absorption	6.6%

CB 1038 Casting Earthenware

Kentucky ball	48.0	Temperature	C/02
Flint	28.0	Shrinkage @ C/02	10.0%
Cornwall stone	20.0	Plastic scale	9.0
China clay	4.0	Color	Creamy white (30)
	100.0	Absorption	18.5%
Sodium silicate, dry	0.3		

CB 1039 Creamy Earthenware

Flint	40	Temperature	C/02
Kentucky ball #4	35	Shrinkage @ C/02	6.0%
Kaolin	10	Plastic scale	9.3
Cornwall stone	15	Color	Creamy white (30)
	100	Absorption	9.0%

CB 1040 Earthenware IV

Kentucky ball #4	40	Temperature	C/02–2
China clay	26	Shrinkage @ C/02	9.0%
Flint	16	Plastic scale	9.3
Cornwall stone	18	Color	Off-white (31)
	100	Absorption	18.5%

CB 1041 EPK Earthenware

E.P. kaolin	35	Temperature	C/02
Soda feldspar	35	Shrinkage @ C/02	11.0%
Flint	27	Plastic scale	2.0
Bentonite	3	Color	Pinkish-white (63)
	100	Absorption	14.1%

CB 1042 Casting Earthenware

Ball clay	30.0	Temperature	C/02–2
Bone ash	30.0	Shrinkage @ C/02	7.0%
Talc	20.0	Plastic scale	1.5
Silica	20.0	Color	Very light ochre (21)
	100.0	Absorption	21.1%
Sodium silicate	0.2		

Mid-Temperature Clay Bodies

Clay bodies that mature in the C/02-to-C/4 temperature range, between that of earthenware and that of stoneware, can be classified as *mid-temperature* or *medium-temperature*. This in-between-temperature clay is also sometimes referred to as *hard earthenware, soft stoneware,* and *calilloutages* (the French term for a hard, white-colored, high-temperature earthenware containing silica). In general, mid-temperature bodies resemble stoneware more than earthenware, for, like stoneware, they emit a slight ring when struck, they are usually waterproof when unglazed, and their percentage of absorption is lower than that of earthenware. Most of the mid-temperature bodies are semivitrified.

Because of the higher cost of fuel, many potters are now using clay bodies and glazes in the mid-temperature range instead of those in the higher, or stoneware, range. A 10-percent fuel savings can be realized by firing at C/4 rather than at C/10. A word of caution should be given, however. At the lower temperature, the iron spots will be smaller and less pronounced. The addition of granular ilmenite iron, iron chromate, or manganese into the clay body will produce iron spotting.

The following formulas are for throwing, hand-building, and casting bodies:

MID-TEMPERATURE CLAY BODY FORMULAS

CB 1043 Oaktag

Kentucky old mine #4	65	Temperature	C/2
Talc	15	Shrinkage @ C/2	15.0%
Nepheline syenite	15	Plastic scale	4.0
Flint	5	Color	Oaktag (64)
	100	Absorption	5.8%

CB 1044 Sculpture Clay

E.P. kaolin	30	Temperature	C/2
Kentucky ball #4	30	Shrinkage @ C/2	10.0%
Talc	10	Plastic scale	3.0
Cornwall stone	10	Color	Oaktag (30)
Silica sand	10	Absorption	9.7%
Medium grog	10		
	100		

CB 1045 Terra-cotta

Fine grog	25	Temperature	C/2–4
Medium grog	25	Shrinkage @ C/2	15.0%
Ball clay	20	Plastic scale	2.0
Local red clay	20	Color	Dark brick orange (80)
Bentonite	6	Absorption	12.6%
Red iron oxide	4		
	100		

CB 1046 Sculpture Terra-cotta

Fine grog	25	Temperature	C/2–4
Medium grog	25	Shrinkage @ C/2	7.0%
Kentucky ball #4	20	Plastic scale	2.0
Fire clay	20	Color	Brick brown (75)
Red iron oxide	6	Absorption	13.2%
Barium carbonate	4		
	100		

CB 1047 Casting Clay

E.P. kaolin	25	Temperature	C/2
Nepheline syenite	25	Shrinkage @ C/2	15.0%
Ball clay	20	Plastic scale	4.0
Silica	10	Color	Navajo white (31)
Talc	20	Absorption	0.2%
	100		

CB 1048 Oaktag Kaolin

E.P. kaolin	25	Temperature	C/2
Ball clay	25	Shrinkage @ C/2	12.0%
Talc	20	Plastic scale	8.0
Fine grog	15	Color	Oaktag (64)
Cornwall stone	15	Absorption	7.2%
	100		

CB 1049 Navajo White Clay

Kentucky ball #4	65	Temperature	C/2–4
Talc	15	Shrinkage @ C/2	10.0%
Nepheline syenite	15	Plastic scale	9.3
Flint	5	Color	Navajo white (64)
	100	Absorption	12.1%

CB 1050 Eggshell White Clay

Nepheline syenite	50	Temperature	C/2–5
Ball clay	20	Shrinkage @ C/5	10.0%
E.P. kaolin	20	Plastic scale	4.0
Silica	10	Color	Eggshell white (6)
	100	Absorption	8.5%

CB 1051 Dark Brown Sculpture Clay

Redart	40	Temperature	C/2
Goldart	40	Shrinkage @ C/2	12.0%
Grog, red or brick	10	Plastic scale	2.0
Silica sand	5	Color	Dark brown (74)
Granular manganese		Absorption	4.3%
dioxide	3		
Bentonite	2		
	100		

CB 1052 Clay for Making Saggers

Plastic fire clay	40	Temperature	C/2–4
Calcined fire clay	30	Shrinkage @ C/2	12.0%
Fine grog	28	Plastic scale	8.5
Bentonite	2	Color	Light tan (40)
	100	Absorption	18%

CB 1053 Light Tan Clay

Plastic fire clay	40	Temperature	C/2–4
Cedar Heights clay	25	Shrinkage @ C/2	9.5%
Ball clay	15	Plastic scale	2.0
Grog	15	Color	Light tan (24)
Nepheline syenite	5	Absorption	17.6%
	100		
Screened walnut shells	15		

CB 1054 Light Coral Clay

Kentucky ball #4	30	Temperature	C/2–4
Plastic fire clay	30	Shrinkage @ C/2	10.0%
Cedar Heights red or		Plastic scale	8.0
other red clay	15	Color	Light coral (59)
Fine grog	15	Absorption	13.9%
Cornwall stone	10		
	100		

CB 1055 Kentucky Oaktag Clay

Kentucky ball #4	65	Temperature	C/3–5
Talc	20	Shrinkage @ C/3	10.0%
Flint	10	Plastic scale	9.5
Nepheline syenite	5	Color	Oaktag (30)
	100	Absorption	12.9%

CB 1056 Sculpture Clay

Plastic fire clay	60	Temperature	C/3–4
Fine grog	20	Shrinkage @ C/3	10.0%
Ball clay	10	Plastic scale	8.0
Wollastonite	10	Color	Light tan (24)
	100	Absorption	11.2%

CB 1057 Salt-Firing Clay

Kentucky old mine #4	60	Temperature	C/3–4
Flint	20	Shrinkage @ C/3	10.0%
Soda feldspar	10	Plastic scale	9.5
Fire clay	5	Color	Oaktag (64)
Talc	5	Absorption	14.7%
	100		

CB 1058 Slipcasting Clay

Nepheline syenite	50.0	Temperature	C/3–5
Kentucky ball	35.0	Shrinkage @ C/3	9.0%
Talc	15.0	Plastic scale	8.5
	100.0	Color	Oaktag (31)
Sodium silicate, dry	0.2	Absorption	11.8%

CB 1059 Delft Blue Casting Clay

Cornwall stone	45.5	Temperature	C/3
Kentucky ball	23.2	Shrinkage @ C/3	10.0%
China clay	22.2	Plastic scale	1.5
Cobalt oxide	9.1	Color	Delft blue (91)
	100.0	Absorption	9.8%

CB 1060 Pink Brick Terra-cotta

Ball clay	25	Temperature	C/3–5
Brick grog	20	Shrinkage @ C/3	10.0%
Grog, mixed sizes	20	Plastic scale	9.0
Plastic fire clay	20	Color	Pink brick (59)
Nepheline syenite	5	Absorption	12.7%
Bentonite	5		
Talc	5		
	100		

CB 1061 Oaktag Clay

Ball clay	61.1	Temperature	C/4–6
China clay	15.3	Shrinkage @ C/4	11.0%
Flint	13.9	Plastic scale	9.3
Cornwall stone	9.7	Color	Oaktag (30)
	100.0	Absorption	13.3%

CB 1062 Light Pink Casting Clay

Custer feldspar	60	Temperature	C/4
Flint	18	Shrinkage @ C/4	5.0%
Cornwall stone	16	Plastic scale	2.0
China clay	6	Color	Light pink (63)
	100	Absorption	8.5%

CB 1063 Rockingham Casting Clay

Ball clay	48.7	Temperature	C/4–8
China clay	31.7	Shrinkage @ C/6	14.0%
Flint	17.0	Plastic scale	4.0
E.P. kaolin	2.6	Color	Oaktag (15)
	100.0	Absorption	13.1%

CB 1064 Bone White Clay

Ball clay	40	Temperature	C/4–5
E.P. kaolin	30	Shrinkage @ C/3	8.0%
Flint	15	Plastic scale	9.0
Soda feldspar	15	Color	Bone (6)
	100	Absorption	16.5%
Bentonite	2		

CB 1065 Les Lawrence's Clay Body

Kentucky ball	36.4	Temperature	C/4–5
Lincoln fire clay	36.4	Shrinkage @ C/4	11.0%
Silica sand	18.1	Plastic scale	9.3
Plastic vitrox	9.1	Color	Light tan (24)
	100.0	Absorption	9.8%

Talcware

Talc—also known as *magnesium silicate, French chalk, steatite,* and *soapstone*—is an inexpensive source of magnesia and silica. In a clay body, the magnesia acts as a flux at high temperatures, while talc improves resistance to thermal shock and to acid attack. In low-temperature bodies, talc increases thermal expansion. In stoneware bodies, talc serves as the flux, reducing the amount of feldspar needed to produce the desired strength; it also decreases the crazing of glazes. In wall-tile bodies in which talc is used in amounts up to 50 percent, no feldspar is added; where talc is used in amounts up to 10 percent, only a small amount of feldspar is used. High-talc wall-tile bodies have lower shrinkage, greater strength, and a much higher resistance to crazing than typical clay-silica-feldspar bodies have. Very-high-talc wall bodies—those containing 60 to 70 percent talc—are fired in the C/03–2 range. High-percentage-talc wall bodies are popular because they have almost no crazing, a low firing-temperature range, fast firing properties, and a perfectly white body which enhances brightly colored glazes. In casting bodies on which alkaline glaze is applied (such as C/06–5 bodies), high amounts of talc are used to prevent crazing and shivering.

It should be noted that because talc is mined in a number of locations and may vary greatly according to location of mine, the results obtained with one talc cannot be assumed to apply to another. Generally, lime-bearing talc is desirable for casting and dinnerware bodies.

The following talc body formulas include those for casting and throwing.

TALC BODY FORMULAS

CB 1066 Talcware I

Ball clay	55	Temperature	C/012–02
Talc	35	Shrinkage @ C/012	5.0%
Gerstley borate	10	Plastic scale	8.0
	100	Color	Pinkish oaktag (63)
		Absorption	15.2%

CB 1067 Hand-building Talcware

Talc	50	Temperature	C/06–02
Ball clay	50	Shrinkage @ C/06	7.0%
	100	Plastic scale	9.3
Silica sand	30	Color	Off-white (64)
		Absorption	10.9%

CB 1068 Pinkish-White Talcware

Talc	50	Temperature	C/06–04
Ball clay	33	Shrinkage @ C/06	7.0%
Plastic vitrox	17	Plastic scale	8.5
	100	Color	Pinkish-white (63)
		Absorption	13.4%

CB 1069 Whiteware Oxidation Talcware

Talc	44	Temperature	C/06–02
Ball clay	44	Shrinkage @ C/06	9.0%
Silica	8	Plastic scale	7.0
Whiting	4	Color	Creamy white (6)
	100	Absorption	13.1%

CB 1070 Whiteware Slipcasting Talcware

Talc	38	Temperature	C/06–1
Ball clay	38	Shrinkage @ C/06	8.0%
Silica	24	Plastic scale	Not measured
	100	Color	Oaktag (30)
		Absorption	13.8%

CB 1071 Slipcasting Talcware

Talc	62.0	Temperature	C/04–02
Ball clay	33.0	Shrinkage @ C/04	7.0%
Whiting	5.0	Plastic scale	Not measured
	100.0	Color	Creamy white (6)
Barium carbonate	0.5	Absorption	15.2%
Sodium silicate	0.2		

CB 1072 Talcware II

Ball clay	50	Temperature	C/04–2
Talc	40	Shrinkage @ C/04	10.0%
Frit #2106	10	Plastic scale	9.0
	100	Color	Oaktag (64)
		Absorption	19.2%

CB 1073 Talcware III

Kentucky ball #4	50	Temperature	C/04–2
Talc	40	Shrinkage @ C/04	10.0%
Nepheline syenite	10	Plastic scale	9.5
	100	Color	Oaktag (30)
		Absorption	19.1%

CB 1074 Talcware IV

Ball clay	50	Temperature	C/04–2
Talc	30	Shrinkage @ C/04	10.0%
Plastic fire clay	20	Plastic scale	9.3
	100	Color	Oaktag (64)
		Absorption	18.5%

CB 1075 Casting Talcware

Kentucky ball	40.0	Temperature	C/04–02
Talc	40.0	Shrinkage @ C/04	12.0%
E.P. kaolin	10.0	Plastic scale	8.0
Gerstley borate	10.0	Color	Oaktag (15)
	100.0	Absorption	8.7%
Sodium silicate, dry	0.3		

Stoneware

Stoneware is fired in the C/4–11 range and has a maturing temperature between that of earthenware and of porcelain. It is a popular clay body known for its strength, earth-tone colors, hardness, and vitrification. Like porcelain, it is fired to the point where it is vitrified and is thus impervious to liquids. Unlike porcelain, however, it is seldom more than faintly translucent. Its glaze and body mature at the same temperature to form an integrated glaze fit.

In the Middle Ages, lead-glazed earthenware pottery was common in Europe. By the twelfth century, German potters had produced a high-temperature earthenware, a proto-stoneware to replace Pingsdorf pottery of the Rhineland. During the late fourteenth century, German earthenware emerged as a true stoneware, the best-known example of which was the salt-glazed pottery made in the Siegburg region. Later (about 1545), the French Beauvais potters developed their own nonporous pottery. John Dwight (late seventeenth century) introduced stoneware into England. There it was further refined by the Eler brothers, who utilized techniques from Germany. Important types of stoneware developed in Europe in the 1600s and 1700s include *basalt* (also called *Egyptian stoneware*), which uses cobalt and manganese oxides to make a black stoneware; *agateware* (also called *marbleware* or *variegated ware*), which is an imitation of agatestone and is produced by combining different-colored clays in a single body; and *jasper*, a hard, fine-grained, unglazed, and slightly translucent stoneware. Introduced by Josiah Wedgwood (about 1774), jasper is colored with oxides and stains to produce blue, sage-green, yellow, black, and lilac hues. Its most common ceramic form has been relief decoration depicting Greco-Roman scenes.

The stoneware of the past was usually produced from natural

clays containing feldspar and silica, which fired to a dense and nonporous state. Several of the natural stoneware clays are available today, including Zanesville, Jordan, Monmouth, and Cedar Heights. Other natural clays, such as fire clay, sewer pipe, and some brick clays, can be used directly, without any additives, and are acceptable for throwing and hand building. Most commercially prepared clays are compounded from various clays (ball, fire, and china), feldspar, and silica in the proportions necessary to obtain the desired properties. The fired colors are gray, buff, tan, brown, and orange. Stoneware clays generally contain iron, which distinguishes them from porcelain (it is the lack of iron, in fact, that gives porcelain its characteristic white color and translucency).

The desired attributes of stoneware include plasticity, with a wide workability range; strength but not brittleness; minimal shrinkage; earthy colors; iron spotting; minimal warping; resistance to bloating; and proper fusing of glaze and clay. The natural iron in the clay, or iron added to the clay (as iron, or as ilmenite or manganese), migrates, when fired, to the surface of the clay and to the glaze, producing the characteristic iron-spotting effect.

Reduction firing is one of the major characteristics of stoneware ceramics. The purpose of reduction is to remove oxygen from the metallic oxides in the clay and glaze during the firing process and thereby produce a wide range of colors or particular surface effects. In gas firing the atmosphere can be *oxidative* (more oxygen is in the kiln than can be used by the flame), *neutral* (a balanced flame), or *reductive* (less oxygen than needed by the flame); in electric firing it is always neutral. To obtain reduction in gas firing, the operator controls the quantity of gas and air (both primary and secondary) going into the kiln and the dampers, by decreasing the amount of exhaust. In an electric kiln, the operator introduces burnable organic materials, such as dry wood slivers, mothballs, and wood shavings. The excess carbon (smoke) in the kiln is chemically active and will readily combine with the oxygen of the metallic oxides to form carbon monoxide and carbon dioxide (which will escape out the exhaust). Some metallic oxides—like copper and iron oxides, which show the most visible changes—will readily give up their oxygen atoms. Depending upon the makeup of the glaze, the amount of oxide, the temperature fired, and the kiln atmosphere, copper oxide can produce blood red, green, blue, gray, black, and shades of pink; iron oxide, under the same conditions, can produce black, brown, burnt orange, rust, green, celadon, and tan. Other metallic oxides do not show as dramatic a change. At lower temperatures, the metals tin, gold, bismuth, silver, platinum, and copper will deposit a thin layer of metal on or in the glaze, creating luster and mother-of-pearl effects. Should the reduction be very heavy, however, many glazes containing 10 percent or more metallic oxide (especially manganese oxide) will bloat in the clay body and blister the glaze.

The most popular stoneware firing temperatures are C/5–6, C/8, and C/9–10, although this will vary with the desires of the individual ceramist. Even though there may be little significant difference in the temperatures (10° or less), one ceramist will choose to fire to a light C/10, while another will prefer to fire to a heavy C/9. Temperatures can also vary according to the location of the cones in the kiln. Those placed in front of the spy hole will be cooler than those placed at the back of the kiln. Other factors, as well, will determine the appropriate temperature. A C/9 firing with a long soaking, for example, will produce results similar to those of a C/10 firing with no soaking. The variations go on endlessly! Further, glazing results will also vary, depending on the thickness of the glaze, accurate measurement of the formula, the clay body and local glaze material used, firing temperature, the character of the kiln atmosphere (the amount of oxygen in the kiln chamber, a condition determined by kiln design and by the method of heat production), the length of the soaking period, and the whim of the "kiln gods." Two potters, using the same glaze formula, will often get slightly different results. Therefore, glaze testing is highly advisable.

The following clay formulas are divided into three divisions: C/5–6, C/8, and C/9–10.

STONEWARE CLAY BODY FORMULAS

CONE/5–6

CB 1076 Pumpkin Hand-building Stoneware

Greenstripe fire clay	59	Temperature	C/5–8
Wollastonite	15	Shrinkage @ C/5	10.0%
Large grog	9	Plastic scale	6.0
Medium grog	8	Color	Pumpkin (27)
Silica sand	8	Absorption	10.8%
Barium carbonate	1		
	100		

CB 1077 Cornwall Casting Stoneware

Cornwall stone	50.4	Temperature	C/5
China clay	25.2	Shrinkage @ C/5	11.0%
Ball clay	24.4	Plastic scale	4.0
	100.0	Color	Oaktag (30)
		Absorption	5.1%

CB 1078 Casting White Stoneware

China clay	50.0	Temperature	C/5–7
Flint	30.0	Shrinkage @ C/5	10.0%
Soda feldspar	15.0	Plastic scale	Not measured
Bentonite	5.0	Color	Eggshell white (6)
	100.0	Absorption	17.6%
Soda ash	0.3		

CB 1079 Jordon Stoneware

Jordon clay	50	Temperature	C/5
Ione grog, fine mesh	30	Shrinkage @ C/5	11.0%
Red grog	15	Plastic scale	9.0
Bentonite	5	Color	Muted orange brick (80)
	100	Absorption	0.1%

CB 1080 Gray Stoneware

Nepheline syenite	50	Temperature	C/5
Tennessee ball	30	Shrinkage @ C/5	7.0%
E.P. kaolin	15	Plastic scale	8.5
Bentonite	5	Color	Light gray (31)
	100	Absorption	0.1%

CB 1081 Staffordshire Stoneware I

Flint	35.3	Temperature	C/5–8
Ball clay	27.0	Shrinkage @ C/5	7.0%
China clay	22.6	Plastic scale	8.0
Cornwall stone	15.1	Color	Oaktag (15)
	100.0	Absorption	13.5%

CB 1082 Casting Stoneware I

Flint	32.8	Temperature	C/6–7
Ball clay	28.8	Shrinkage @ C/6	8.0%
China clay	27.8	Plastic scale	6.0
Cornwall stone	10.6	Color	Oaktag (64)
	100.0	Absorption	12.2%

CB 1083 Staffordshire Stoneware II

Ball clay	29.1	Temperature	C/6–8
Flint	28.5	Shrinkage @ C/6	15.0%
China clay	27.1	Plastic scale	9.3
Cornwall stone	15.2	Color	Oaktag (30)
	99.9	Absorption	10.3%

CB 1084 Casting Stoneware II

Ball clay	29.0	Temperature	C/5–7
Fine grog	28.0	Shrinkage @ C/5	8.0%
Cornwall stone	22.0	Plastic scale	8.0
China clay	14.0	Color	Oaktag (30)
Flint	7.0	Absorption	11.4%
	100.0		
Sodium silicate (fl. oz.)	0.2		

CB 1085 Fire Clay Stoneware

Fire clay	60	Temperature	C/6
Ball clay	20	Shrinkage @ C/6	15.0%
Sand, 60 mesh	20	Plastic scale	9.0
	100	Color	Light tan (24)
		Absorption	4.3%

CB 1086 Kaolin Casting Stoneware

E.P. kaolin	55	Temperature	C/6–8
Flint	40	Shrinkage @ C/6	13.0%
Kingman feldspar, old	5	Plastic scale	0
	100	Color	Creamy white (6)
		Absorption	13.6%

CB 1087 Dark Ochre Hand-building Stoneware

Bonding clay	47.0	Temperature	C/6
Fine grog	24.6	Shrinkage @ C/6	10.0%
Redart	15.7	Plastic scale	2.5
Granular manganese,		Color	Dark ochre (27)
40–60 mesh	12.3	Absorption	4.3%
Bentonite	0.4		
	100.0		

CB 1088 Brown Stoneware

Calcined kaolin	40	Temperature	C/6–7
Fire clay	30	Shrinkage @ C/6	15.0%
Fine grog	26	Plastic scale	7.5
Red iron oxide	2	Color	Light brown (75)
Bentonite	2	Absorption	7.3%
	100		

CB 1089 Tan Stoneware

Fine grog	40	Temperature	C/6–7
Plastic fire clay	40	Shrinkage @ C/6	15.0%
Ball clay	20	Plastic scale	9.5
	100	Color	Light tan (38)
		Absorption	9.3%

CB 1090 Sandy Stoneware

Plastic fire clay	40	Temperature	C/6–8
Ball clay	20	Shrinkage @ C/6	13.0%
Mixed grog	20	Plastic scale	9.0
Soda feldspar	10	Color	Oaktag (64)
Silica sand	10	Absorption	13.6%
	100		

CB 1091 Gray Hand-building Stoneware

Calcined kaolin	40	Temperature	C/6–7
Mixed grog	40	Shrinkage @ C/6	12.0%
Ball clay	15	Plastic scale	4.0
Bentonite	5	Color	Cement gray
	100	Absorption	8.9%

CB 1092 Light Ochre Stoneware

Plastic fire clay	40	Temperature	C/6
Calcined kaolin	20	Shrinkage @ C/6	15.0%
Ball clay	20	Plastic scale	9.3
Grog	20	Color	Light ochre (29)
	100	Absorption	7%

CB 1093 German Stoneware

Plastic fire clay	36	Temperature	C/6–8
E.P. kaolin	30	Shrinkage @ C/6	12.0%
Flint	30	Plastic scale	8.0
Kingman feldspar	4	Color	Marbled light tans (62)
	100	Absorption	11.7%

CB 1094 Speckled Stoneware

Ball clay	32	Temperature	C/6
Plastic vitrox	32	Shrinkage @ C/6	11.0%
Mixed grog	20	Plastic scale	8.0
Talc	8	Color	Light speckled
Bentonite	8		gray (151)
	100	Absorption	0.1%

CB 1095 Dense Stoneware

E.P. kaolin	30	Temperature	C/6
Ball clay	30	Shrinkage @ C/6	13.0%
Kingman feldspar	30	Plastic scale	8.3
Silica	10	Color	Oaktag (30)
	100	Absorption	3.8%
Silica sand	20		

CB 1096 Casting Stoneware

Georgia kaolin	27.0	Temperature	C/6
Soda feldspar	20.0	Shrinkage @ C/6	11.0%
Silica	18.0	Plastic scale	9.3
Jordon clay	18.0	Color	Light tan (24)
Kentucky ball	17.0	Absorption	2.9%
Soda ash	0.3		
	100.3		

CB 1097 Stoneware

Calcined kaolin	25	Temperature	C/6–7
Calcined stoneware	25	Shrinkage @ C/6	14.0%
Ione grain grog	15	Plastic scale	8.5
Tennessee ball	15	Color	Light ochre-tan (28)
Greenstripe fire clay	10	Absorption	8.0%
Bentonite	10		
	100		

CB 1098 Casting Semiporcelain Stoneware

Bone ash	25.0	Temperature	C/6
E.P. kaolin	25.0	Shrinkage @ C/6	12.0%
Ball clay	25.0	Plastic scale	0
Potash feldspar	15.0	Color	Creamy white (6)
Flint	10.0	Absorption	0.1%
	100.0		
Sodium silicate, dry	0.3		

CB 1099 Speckled Brown Hand-building Stoneware

Fine grog	20	Temperature	C/6–8
Medium grog, red	20	Shrinkage @ C/6	14.0%
E.P. kaolin	20	Plastic scale	4.0
Fire clay	20	Color	Medium speckled
Sawdust or walnut			brown (74)
shells, screened	10	Absorption	9.2%
Bentonite	3		
Red iron oxide	3		
Talc	4		
	100		

CONE/8

CB 1100 Manchester's Stoneware

Cedar Heights goldart	69.0	Temperature	C/8
Tennessee ball	13.8	Shrinkage @ C/8	15.0%
Missouri greenstripe	6.9	Plastic scale	9.5
Silica sand	6.9	Color	Light tan (28)
Redart	3.4	Absorption	6.7%
	100.0		

CB 1101 Sewer Pipe Clay Stoneware

Sewer pipe clay	60	Temperature	C/8
E.P. kaolin	20	Shrinkage @ C/8	16.0%
Grog or silica sand	10	Plastic scale	4.0
Ball clay	10	Color	Medium pinkish-
	100		brown (37)
		Absorption	4.2%

CB 1102 Jordon Stoneware

Jordan clay	55	Temperature	C/8–9
Ball clay	22	Shrinkage @ C/8	15.0%
Flint	18	Plastic scale	8.5
Buckingham feldspar	5	Color	Ochre-orange
	100	Absorption	7.2%

CB 1103 Stoneware

E.P. kaolin	55	Temperature	C/8
Potash feldspar	25	Shrinkage @ C/8	18.0%
Flint	15	Plastic scale	9.0
Bentonite	5	Color	Creamy white (6)
	100	Absorption	4.4%

CB 1104 Goldart Stoneware

Goldart	46	Temperature	C/8–9
Fire clay	32	Shrinkage @ C/8	14.0%
Ball clay	15	Plastic scale	9.5
Sand, 60 mesh	6	Color	Light tan (29)
Bentonite	1	Absorption	5.3%
	100		

CB 1105 Bone Ash Casting Stoneware

Bone ash	45	Temperature	C/8–10
China clay	25	Shrinkage @ C/8	14.0%
Kingman feldspar	20	Plastic scale	1.5
E.P. kaolin	10	Color	Bone white
	100	Absorption	21.6%

CB 1106 Casting Stoneware

Potash feldspar	45.0	Temperature	C/8
E.P. kaolin	35.0	Shrinkage @ C/8	18.0%
Flint	20.0	Plastic scale	0
	100.0	Color	White (32)
Sodium silicate, dry	0.3	Absorption	5.4%

CB 1107 Orange Stoneware

Ball clay	42.4	Temperature	C/8–9
China clay	30.3	Shrinkage @ C/8	16.0%
Fire clay	24.2	Plastic scale	9.0
Black iron oxide	3.1	Color	Medium orange
	100.0	Absorption	6.0%

CB 1108 Ochre Stoneware

Red clay	40.0	Temperature	C/8–9
China clay	25.0	Shrinkage @ C/8	12.0%
Fire clay	22.5	Plastic scale	2.0
Flint	12.5	Color	Marbled ochre and
	100.0		medium orange
		Absorption	9.2%

CB 1109 Whitish Stoneware

Talc	33	Temperature	C/8
E.P. kaolin	33	Shrinkage @ C/8	18.0%
Kentucky ball #4	23	Plastic scale	8.5
Soda feldspar	7	Color	Light grayish
Bentonite	4		white (31)
	100	Absorption	2.3%

CB 1110 Creamy White Casting Stoneware

Flint	30	Temperature	C/8–9
Nepheline syenite	30	Shrinkage @ C/8	12.0%
E.P. kaolin	25	Plastic scale	0
Tennessee ball	15	Color	Creamy white (6)
	100	Absorption	8.8%
		Note: Good for crystal glazes	

CB 1111 Off-White Stoneware

Kaolin	28	Temperature	C/8–9
Flint	28	Shrinkage @ C/8	16.0%
Nepheline syenite	23	Plastic scale	7.0
Ball clay	19	Color	Oaktag (6)
Bentonite	2	Absorption	7.4%
	100		

CONE/9–10

CB 1112 Mouse Gray Stoneware

Lincoln clay	71	Temperature	C/9
Monmouth clay	18	Shrinkage @ C/9	14.0%
Buckingham feldspar	7	Plastic scale	9.5
Flint	4	Color	Mouse gray
	100	Absorption	0.1%

CB 1113 Brown Casting Stoneware

Lincoln clay	65	Temperature	C/9
Monmouth clay	15	Shrinkage @ C/9	14.0%
Buckingham feldspar	10	Plastic scale	2.0
Flint	10	Color	Cardboard brown (34)
	100	Absorption	0.1%

CB 1114 Brick Brown Stoneware

Greenstripe fire clay	55.5	Temperature	C/9
Ball clay	27.8	Shrinkage @ C/9	25.0%
Sand, 30–60 mesh	11.1	Plastic scale	9.5
Kingman feldspar	5.6	Color	Brick brown (73)
	100.0	Absorption	0.1%
		Note: Throws well, very coarse	

CB 1115 Fireclay Stoneware

Fire clay	55	Temperature	C/9
Kentucky ball #4	26	Shrinkage @ C/9	15.0%
Kona F-4 feldspar	10	Plastic scale	9.5
Granular red iron	4	Color	Medium purplish
Talc	3		brown (146)
Bentonite	2	Absorption	0.1%
	100		

CB 1116 Kaolin Casting Stoneware

E.P. kaolin	50	Temperature	C/9
Kona F-4 feldspar	20	Shrinkage @ C/9	19.0%
Flint	12	Plastic scale	2.0
Fire clay	12	Color	Very pale gray (7)
Silica sand	4	Absorption	0.1%
Macaloid	2	Note: Good for salt-firing	
	100		

CB 1117 Light Coral Casting Stoneware

Fire clay	50	Temperature	C/9
Flint	20	Shrinkage @ C/9	11.0%
Feldspar	10	Plastic scale	2.0
Zircopax	10	Color	Light coral (40)
Fine grog	10	Absorption	2.8%
	100		

CB 1118 Stoneware

Plastic fire clay	50	Temperature	C/9
Fire clay	20	Shrinkage @ C/9	16.0%
Soda feldspar	10	Plastic scale	9.5
Fine grog	5	Color	Cardboard brown (34)
Talc	5	Absorption	0.1%
Silica	10	Note: Good for salt-firing	
	100		

CB 1119 Lincoln Stoneware

Lincoln fire clay	50	Temperature	C/9
Ball clay	25	Shrinkage @ C/9	13.0%
Sand, 30–60 mesh	20	Plastic scale	9.0
Kingman feldspar	5	Color	Light grayish
	100		brown (36)
		Absorption	2.4%

CB 1120 Honest John's Super Stoneware I

Lincoln fire clay	42	Temperature	C/9
Kentucky ball #4	33	Shrinkage @ C/9	15.0%
Del Monte feldspar	25	Plastic scale	7.0
	100	Color	Light gray (151)
Fine grog	10	Absorption	0.1%

CB 1121 Honest John's Super Stoneware II

Lincoln fire clay	40	Temperature	C/9
Kentucky ball #4	40	Shrinkage @ C/9	15.0%
Del Monte feldspar	20	Plastic scale	8.5
	100	Color	Light pumpkin (39)
		Absorption	0.1%

CB 1122 Brick Orange Stoneware

A. P. Green fire clay	40	Temperature	C/9–10
Jordon stoneware	25	Shrinkage @ C/9	15.0%
Kentucky old mine #4	10	Plastic scale	9.5
Soda feldspar	10	Color	Brick orange (80)
Silica	10	Absorption	3.3%
Fine grog	5		
	100		

CB 1123 Rock Stoneware

E.P. kaolin	39	Temperature	C/9
Kentucky ball	30	Shrinkage @ C/9	17.0%
Flint	10	Plastic scale	9.0
Kingman feldspar	10	Color	Light gray (151)
Fine grog	11	Absorption	1.5%
	100	Note: Good for salt-firing	

CB 1124 Self-Glazing Stoneware

Flint	30.6	Temperature	C/9–11
Ball clay	30.6	Shrinkage @ C/9	12.0%
Lincoln clay	20.4	Plastic scale	9.5
Fine grog	10.2	Color	Black glaze (39)
Soda ash	3.1	Absorption	9.4%
Sodium bicarbonate	3.1	Note: Glossy surface	
Bentonite	2.0		
	100.0		

CB 1125 Sculptural Stoneware

E.P. kaolin	28	Temperature	C/9–10
Flint	20	Shrinkage @ C/9	15.0%
Kentucky ball #4	15	Plastic scale	9.0
Potash feldspar	15	Color	Grayish oaktag (30)
Silica sand, 60 mesh	10	Absorption	5.3%
Grog, 30 mesh	10		
Bentonite	2		
	100		

Ovenware

For the ceramist making heat-resistant, thermal-shock-proof cookware products—such as casseroles and roasting pans, pie plates, and frying pans—there are two types of clay bodies available: *flameware* and *ovenware*. Both flameware and ovenware products can be placed in a hot oven and can sustain the thermal shock of being moved directly from a hot oven to the dining room table. Flameware can withstand greater *thermal differential* than ovenware can, however. For example, a cookware product made of flameware can undergo "spot heating," in which a flame heats only one part of a pot for a prolonged period. Products made of flameware, moreover, can be placed directly on a gas range burner or on an electric burner (with wire spacer). Clay bodies made of flameware, likewise, differ from those made of ovenware. For a clay body to be called flameware, it must be able to withstand sudden and drastic thermal shock—such as the quenching of a 200°-hot clay body in cold water. Flameware must also have a smooth, hard finish and possess reasonably good working and mechanical properties.

The key to successful cooking with ceramic products (including those made from earthenware, stoneware, and porcelain) is careful handling. If used properly, ceramic casseroles, baking dishes, and serving trays can be employed both for baking in the oven and for serving hot foods at table. Refrigerator-cold or room temperature foods—but *not frozen foods*—may be placed in ceramic cooking containers, which, in turn, should be placed in a cold oven in which the heat is raised slowly. The danger of breakage occurs when a full casserole is taken out of the freezer and placed directly in a 400°-oven; the pot is very likely to crack. Proper handling of ceramic cookware is not confined to refrigerator-to-oven use, however; for when working with ceramic cooking, one must exercise care at every step. Slamming an oil-filled ceramic frying pan down on a lit burner; tilting a baking dish and allowing the oil to spill onto an open flame; bumping the ware against a hard surface causing it to break; and picking up a hot casserole with bare hands are examples of some of the careless food-preparation techniques that can lead to disaster. It is primarily because of avoidable accidents, though, that ceramic cookware in general and flameware in particular have received some unwarranted bad publicity. On the other hand, some potters have made clay products and called them flameware when they were not; made poorly designed products with too-small handles, ill-fitting lids, or improper proportions; used poor craftsmanship; and claimed that their products were able to do more than could reasonably be expected. Such practices cannot help but give ceramic cookware a bad name.

Many stoneware bodies too can be made into satisfactory ovenware. For example, a C/10 stoneware can have 5 percent talc and 19 percent grog (ione or other low "free" silica type) wedged in and be fired to C/8. These additives and the lower firing temperature will produce a more open clay body that has less warpage and is better able to withstand thermal shock. Some clay bodies, however, have too much "free" silica, iron, or impurities to be easily converted to ovenware.

Most heat-resistant body formulas are based on *cordierite*, a crystalline material compounded of magnesite (22 percent), kaolin (70 percent), and silica (8 percent); *clay grog,* a coarse body composed of low silica grog (50 percent), kaolin (40 percent), and flux (10 percent); *lithia,* the best known of the flameware compounds, composed of spodumene (40 percent) and ball clay (60 percent); or *talc clay,* which has a very short maturing range and is made up of talc (40 percent), kaolin (40 percent), alumina (10 percent), and flux (10 percent). The development of pure thermal-shock bodies is difficult, for the compounds, cordierite in particular, have short firing ranges of $+/- 5°$F. The addition of a flux such as lead, borate, soda feldspar, potash feldspar, whiting, dolomite, soda, or borax will satisfactorily extend the vitrification range of these bodies. Additions of 10 percent to 30 percent flux will result in the vitrification of a body in the lower temperature ranges.

Most heat-resistant bodies are fired in the C/6 to 13 range. As the body becomes vitrified, it tends to become brittle, glossy, and less resistant to abuse. Therefore, semivitrification is more desirable, because it results in a stronger body. A porous, low-vitrified body (like *raku*), while physically weak, is highly resistant to thermal shock.

The design and craftsmanship of ovenware is critical. Cooking and serving containers should be designed simply, with uniform wall and bottom thickness; small but functional handles; easy-to-clean interior and exterior surfaces; no unnecessary clay decorations; small foot; easy to use, well-attached handles, spouts, and other clay additions; no sharp corners; and slightly curved walls, bottom, and lids (Drawing 2). Any imperfections in ovenware products will readily show up during use.

Flameware production is both difficult and exacting. The composition of the body is crucial—it must be a silica-free compound—and the clay body will need controlled soaking periods, precise firing periods, highly accurate temperature control, and excellent craftsmanship. Because of the many problems involved in flameware production, flameware formulas are not included in this book. Unfortunately, information about flameware production is limited, and the ceramist will probably have to do extensive research. (For further research, consult the references in the Bibliography.)

Besides the traditional tests used on clay bodies, one other test should be performed on ovenware bodies: a test pot should be made

Drawing 2
Recommended Design Features of Ovenware Pottery

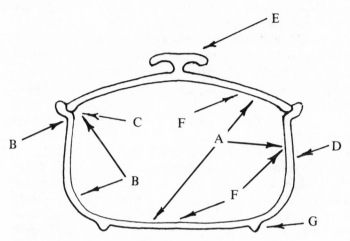

A. Uniform wall and bottom thickness
B. No sharp corners
C. Ease of cleaning and serving
D. No unnecessary clay decoration
E. Small but functional handles
F. Slightly curved walls, bottom, and lid
G. Small foot

of the clay body to be used. A pound of ovenware clay is weighed and thrown into a slightly curved-bottom bowl, about 4 inches wide and 2 inches high, with uniform wall and bottom thickness. The pot is bisque-fired; glaze is applied; and then it is fired to recommended temperature. Next, the finished pot is filled three-quarters full of water, placed on top of a gas burner—*not electric*—or in an oven to heat. After the water evaporates, the pot is heated 15 minutes longer; the heat is then turned off and the pot allowed to cool. This heat-treatment test will determine whether the clay body will be able to withstand the thermal shock of everyday cooking use.

Variations of pure thermal-shock bodies have been given here in order to cover a wider temperature range and to include bodies that can be used on a potter's wheel. The formulas include C/06, 04, 6, 8, and 9/10 maturing temperatures; and white, tan, brown, and gray color bodies. All of these bodies can be wheel-thrown.

OVENWARE CLAY BODY FORMULAS

CB 1126 Navajo White Ovenware

Kentucky ball	40	Temperature	C/06–02
Calcined kaolin	30	Shrinkage @ C/04	10.0%
Talc	20	Plastic scale	9.5
Fine grog	5	Color	Navajo white (64)
Spodumene	5	Absorption	17.1%
	100	Glaze scale	
		Surface	Gloss
		Fluidity	Little
		Stain penetration	All
		Opacity	Transparent
		Color / oxidation	Clear
		Note: Slight cracking	

CB 1127 Old Mine Ovenware

Kentucky old mine #4	45	Temperature	C/04
Talc	26	Shrinkage @ C/04	9.0%
Georgia clay	19	Plastic scale	9.0
Frit #54 PEMCO	10	Color	Oaktag (30)
	100	Absorption	8.0%
		Glaze scale	
		Surface	Semigloss
		Fluidity	None
		Stain penetration	All
		Opacity	Transparent
		Color	Clear
		Note: Fine cracks	

CB 1128 Pumpkin Ovenware

Spodumene	40	Temperature	C/6–10
Plastic fire clay	40	Shrinkage @ C/6	10.0%
Ball clay	20	Plastic scale	9.3
	100	Color	Light pumpkin (61)
		Absorption	14.4%
		Glaze scale	
		Surface	Gloss/semigloss
		Fluidity	None
		Stain penetration	All
		Opacity	Transparent
		Color	Warm tan

CB 1129 Ovenware

Cedar Heights goldart	30	Temperature	C/6
A. P. Green fire clay	30	Shrinkage @ C/6	17.0%
Oxford feldspar	20	Plastic scale	9.3
Tennessee ball	10	Color	Medium gray (8)
Flint	5	Absorption	2.0%
Grog, 40–80 mesh	5	Glaze scale	Not determined
	100		

CB 1130 Light Gray Ovenware

Talc	30	Temperature	C/6–8
China clay	30	Shrinkage @ C/6	14.0%
Kentucky ball	25	Plastic scale	9.3
Lithium carbonate	5	Color	Light gray (8)
Fine grog	5	Absorption	9.4%
Bentonite	5	Glaze scale	Not determined
	100		

CB 1131 Ovenware (Press-Mold Only)

Fine grog	33.3	Temperature	C/8–11
Alumina hydrate	33.3	Shrinkage @ C/8	7.0%
Jordon clay	33.3	Plastic scale	0
	99.9	Color	Light tan (62)
Bentonite	3.0	Absorption	18.7%
		Glaze scale	
		Surface	Gloss
		Fluidity	Little
		Stain penetration	All
		Opacity	Transparent
		Color	Clear

CB 1132 Throwing Ovenware

Spodumene	25	Temperature	C/8–10
Greenstripe fire clay	25	Shrinkage @ C/8	9.0%
Kentucky ball #4	25	Plastic scale	9.0
Talc	16	Color	Ochre (21)
Ione-grain fine grog	7	Absorption	13.1%
Bentonite	2	Glaze scale	Not determined
	100		

CB 1133 Tan Ovenware

A. P. Green fire clay	50	Temperature	C/9
Tennessee ball	14	Shrinkage @ C/9	16.0%
Oxford feldspar	14	Plastic scale	9.5
Goldart	8	Color	Light tan (29)
Talc	8	Absorption	1.8%
Medium grog	6	Glaze scale	Mesa 333 glaze
	100	Surface	Gloss
		Fluidity	Little
		Stain penetration	All
		Opacity	Transparent
		Color / reduction	Clear
		Note: Minute cracks	

CB 1134 Ochre Ovenware

Greenstripe fire clay	45	Temperature	C/8–9
Spodumene	22	Shrinkage @ C/9	12.0%
Kentucky ball #4	20	Plastic scale	9.5
Talc	5	Color	Ochre (21)
Ione-grain fine grog	5	Absorption	6.5%
Bentonite	3	Glaze scale	Mesa 333 glaze
	100	Surface	Thin—matt
			Thick—gloss
		Fluidity	None
		Stain penetration	Most
		Opacity	Transparent
		Color / reduction	Brown where thin
			Gray where thick

CB 1135 Brick Ovenware

Spodumene	30	Temperature	C/9–10
Jordon clay	27	Shrinkage @ C/9	9.0%
Pyrophyllite	20	Plastic scale	9.0
Kentucky ball #4	20	Color	Brick orange (80)
Talc	2	Absorption	7.3%
Bentonite	1	Glaze scale	Mesa 333 glaze
	100	Surface	Semimatt
		Fluidity	None
		Stain penetration	Darks
		Opacity	Translucent
		Color / reduction	Gray-tan

Low-Shrinkage Clay Bodies

Most throwing porcelain and stoneware clay bodies shrink from 11 percent to 20 percent. Throwing a vase that measures 20 inches before firing and remeasuring it after a C/11 firing to find that it is now only 16½ inches is disheartening. Clay for throwing shrinks more than hand-building or sculpture types because it contains more water and because hand-building and sculpture clays contain coarser clays and grog. Reducing the amount of shrinkage in throwing bodies is difficult because of the need for maintaining clay plasticity. A clay body that contains a high percentage of grog and calcined clay will shrink less but will also be less workable. Regular throwing clay can have fire clay, calcined kaolin, and/or grog—from 20 percent to 50 percent—wedged into the clay to reduce the shrinkage and to open up the clay. Firing at a lower temperature will also help reduce shrinkage. Although such measures will help reduce shrinkage in an existing clay body, the results will not be as good as they would be if a low-shrinkage body had been made from scratch.

It is much easier to develop a low-shrinkage body for hand-building clays. Sculptural and large hand-built forms often have walls that are an inch or more thick. Such a mass will need slow drying and firing to prevent warping, cracking, and the building up of stress. A low-shrinkage open body will permit faster drying and less warping, cracking, and shrinkage.

Criteria used to distinguish low-shrinkage bodies from other clay bodies are 7 percent or less total shrinkage, open body, absorption rate of 2 percent to 11 percent, and workability as a hand-building or a sculpture body. In low-shrinkage bodies, as in casting-clay bodies, the addition of an electrolyte will reduce the amount of water needed to obtain a workable consistency. An electrolyte of 0.1 to 0.5 percent soda ash, 0.1 to 0.3 percent sodium silicate, and/or 0.3 to 0.8 percent macaloid can be used to start. Chopped Fiberglas is available in $\frac{1}{16}$-inch, $\frac{1}{4}$-inch, $\frac{3}{8}$-inch, and $\frac{1}{2}$-inch lengths. It is used in amounts of 1 percent to 5 percent as a filler, as a reinforcement, and as a means of increasing the dry strength of bisque and fired clay. Depending upon the thickness of the strand and its composition, most Fiberglas will either burn out or become absorbed in the clay matrix when fired above C/7.

Two to six percent crushed and screened organic materials, such as walnut or other shells, straw, or sawdust, can be added to the moist clay body. These materials will burn out, leaving the clay body more open, lighter in weight, and with less chance of cracking and warping.

The following formulas are for various types of low-shrinkage bodies and can be used for both sculpture and hand building.

LOW-SHRINKAGE CLAY BODY FORMULAS

CB 1136 Sculpture Clay

Sewer pipe clay	24.8	Temperature	C/3–5
Lincoln fire clay	24.8	Shrinkage @ C/3	7.0%
Wollastonite	12.4	Plastic scale	4.0
Ione grain 420	12.4	Color	Brick orange (80)
Ione grain 400	12.4	Absorption	10.5%
Silica sand, 20 mesh	12.4		
Bentonite	0.8		
	100.0		
Chopped Fiberglas wedges in			
Soda ash	0.3		
Sodium silicate	0.2		

CB 1137 Low-Shrinkage Clay I

Calcined kaolin	50.0	Temperature	C/5–6
Fine grog	20.0	Shrinkage @ C/5	7.0%
Potash feldspar	10.0	Plastic scale	1.5
Chopped fiberglass	5.0	Color	Brown brick (74)
Talc	5.0	Absorption	10.3%
Red iron oxide	5.0		
Bentonite	5.0		
	100.0		
Soda ash	0.3		
Sodium silicate	0.1		

CB 1138 Low-Shrinkage Clay II

Calcined kaolin	25.0	Temperature	C/6
Medium grog	20.0	Shrinkage @ C/6	7.0%
Fine grog	20.0	Plastic scale	5.0
Plastic fire clay	20.0	Color	Light tan (27)
Talc	10.0	Absorption	6.2%
Bentonite	5.0		
	100.0		
Soda ash	0.3		
Sodium silicate	0.1		

CB 1139 Low-Shrinkage Clay III

Spodumene	35.0	Temperature	C/8–9
Calcined kaolin	20.0	Shrinkage @ C/8	6.8%
Kentucky ball	20.0	Plastic scale	8.5
E.P. kaolin	20.0	Color	Light ochre (62)
Talc	5.0	Absorption	10.9%
	100.0		
Soda ash	0.2		
Sodium silicate, dry	0.2		

Self-Glazing Clay Bodies

The best known of the self-glazing clay bodies are Egyptian paste and hard porcelain. *Egyptian paste* is a mixture of grog, kaolin, silica, flux (soda ash, sodium bicarbonate, borax, boric acid, and/or salt), and colorants (metallic carbonates with copper, best known for the turquoise blues). Just enough water is added to the mixture to bring the materials to modeling consistency. Objects are made by hand rather than on the wheel and are generally restricted to small pieces like jewelry and sculpture. The soluble fluxes migrate (effloresce) through the nonplastic clay to the surface during the drying stage. Because of the flux deposit on the surface, the ware is not handled. The object is set on stilts and beads on wire for drying and firing. At the maturing temperature, 1500° to 1700°F, the flux fuses with the silica, forming a glassy coating.

Hard porcelain is known for its vitrified, translucent, and white qualities. When hard porcelain is fired to C/10 and higher, the surface is so dense that it takes on a slight sheen, an almost glaze-like quality. The basic formula is 50 percent china clay and 25 percent each of flint and feldspar. The ceramist can produce a third type of self-glazing clay. Self-glazing stoneware is made by adding soluble fluxes to an open-bodied stoneware clay. As in Egyptian paste, the soluble fluxes, when fired, migrate to the clay surface, which becomes as glassy as porcelain. Colorants may be added to the clay body, or a thin wash may be painted over the surface to provide color. The same hand construction and throwing techniques of regular clay bodies are used except that the trimming is done when the clay is just strong enough to support itself, rather than when it is leather-hard, and the ware is not handled during the drying stage so that the thin film of soluble fluxes will not be brushed off. From 5 percent to 15 percent soluble fluxes can be added to a clay body to produce the self-glazing effect. If too much flux is added, the maturing temperature will be lowered to the point where the clay will sag or bloat.

The following self-glazing C/012 or C/6 formulas are for throwing or hand building:

SELF-GLAZING CLAY BODY FORMULAS

CB 1140 Sutter Tan

Sutter clay	54	Temperature	C/6
Flint	30	Shrinkage @ C/6	11.0%
Colemanite	10	Plastic scale	9.5
Soda ash	2	Color	Light tan (28)
Sodium bicarbonate	2	Absorption	3.1%
Bentonite	2	Surface	Semigloss
	100		

CB 1141 Jordon Self-Glazing

Jordon clay	40	Temperature	C/6
Greenstripe fire clay	20	Shrinkage @ C/6	15.0%
Silica	20	Plastic scale	9.5
Kentucky ball #4	8	Color	Dark brown (153)
Fine grog	5	Absorption	3.0%
Soda ash	5	Surface	Semigloss
Manganese dioxide	2		
	100		

CB 1142 Tan Self-Glazing

Silica	30	Temperature	C/6
Tennessee ball #1	30	Shrinkage @ C/6	12.0%
Greenstripe fire clay	20	Plastic scale	9.5
Gerstley borate	8	Color	Light tan (151)
Sodium bicarbonate	7	Absorption	1.5%
Talc	5	Surface	Semigloss
	100		

CB 1143 Light Gray Self-Glazing

Tennessee ball #1	30	Temperature	C/6
Silica	20	Shrinkage @ C/6	11.0%
Greenstripe fire clay	20	Plastic scale	9.3
Talc	12	Color	Light gray (151)
E.P. kaolin	10	Absorption	4.0%
Fine grog	10	Surface	Semigloss
Sodium bicarbonate	5		
Soda ash	3		
	110		

CB 1144 Gray Self-Glazing

Silica	30	Temperature	C/6
Tennessee ball #1	30	Shrinkage @ C/6	12.0%
Greenstripe fire clay	10	Plastic scale	9.0
Talc	10	Color	Light gray (151)
E.P. kaolin	7	Absorption	5.9%
Fine grog	5	Surface	Semigloss
Soda ash	4		
Sodium bicarbonate	4		
	100		

CB 1145 Oaktag Self-Glazing

Ball clay	53.0	Temperature	C/012
Frit #14		Shrinkage @ C/012	10.0%
(leadless)	47.0	Plastic scale	8.0
	100.0	Color	Oaktag (30)
Sodium silicate, dry	0.3	Absorption	7.0%
		Surface	Slight sheen

Porcelain

The term *porcelain* is derived from the French word *porcelaine*—meaning cowry shell—because of the similarity of that clayware to the cowry shell. Both the clay and the shell are white, smooth, and translucent.

Since prehistoric times humans have known how to make clay objects and harden them in a fire. But it took a long time—probably thousands of years—for humans to develop a food and liquid container that would be simple to construct, made of readily available materials, easy to clean, nonporous (nonabsorbent), and attractive, and which would, at the same time, be a low heat conductor and not impart undesired flavors to the food or drink it contained (neither metal, wood, or earthenware containers meet all these qualifications). Porcelain satisfies these requirements and has gained worldwide popularity as a result.

It is believed that porcelain was first made by the Chinese as early as the Sui Dynasty (c. A.D. 581–617). The excavations at Samarra, in Mesopotamia, have shown, at least, that porcelain was in use before A.D. 800. Since porcelain proved to be an ideal medium for tableware and decorative items, its use spread throughout China, then to Japan; after 1497 (following the discovery of the sea route around the Cape of Good Hope) it became much sought after in Europe. Potteries on the continent tried to solve the mystery of porcelain composition so that they too could manufacture it. In Germany, Holland, Italy, and France, various forms of glazed earthenware were made in imitation of the Chinese substance. A pseudo-porcelain was created in Florence in limited quantities. In 1675, a proto–soft porcelain was produced in France, and developed into a workable soft-paste porcelain at Saint-Cloud in 1696; the factory at Saint-Cloud remained in operation for the next eighty years. Later, other factories emerged, the most successful of them being in Vincennes (about 1738), which was supplanted by the Sèvres factory in 1759. During this time, too, an ever-widening palette of colors was being created. By the late eighteenth century, hard-paste-porcelain production was well under way in Europe.

It was the Western patrons of the Chinese porcelains who stimulated the development of similar ceramic products in Europe. Because the whims of the aristocracy were catered to by ceramists and painters who, working together, created many beautiful and highly decorative pieces, Chinese and Japanese landscapes and other representations came to dominate European fashion. The Baroque, Regency, and Rococo periods also influenced porcelain design, which sought to capture in ceramic the flowers, fruits, angelic figures, cupids, light and dark colors, and curved lines characteristic of those styles. Since the eighteenth century, the style and decoration of porcelain has varied endlessly. Porcelain has served as a "canvas" for miniature pictorial paintings; its designs have reflected Egyptian, Greek, and Roman styles; and today its decoration embraces the modern precept that form follows function.

The basic formula for porcelain is 50 percent china clay and 25 percent each of feldspar and silica. The type of clay and the use to which the body is to be put will determine whether the formula requires substitution or alteration. Most china clays are not plastic, so ball clays and/or bentonite are frequently substituted. Since some ball clays and bentonites contain iron that will discolor the whiteness of the clay body, white bentonite and low-iron ball clay is recommended. To give more body for hand building or throwing, iron-free silica sand or porcelain grog is added to the clay mixture.

One type of porcelain, bone china, is composed of bone ash (45 percent), china clay (30 percent), and flux (25 percent). Bone ash is calcined bones and contains 70 percent calcium phosphate, calcium carbonate, and other similar substances. Bone china is used extensively in England and is renowned for its superior whiteness and translucency. However, bone china has the difficult working property of being nonplastic and thus most of the ware is made by either casting or jiggering techniques. Soft-paste porcelain, also known as *artificial porcelain, fritted porcelain,* and *pâte tendre,* represents an attempt to imitate oriental porcelain. Artificial porcelain is made by adding fluxes to a whiteware body and then firing it to 2000°F. This clay body can be readily cut with a file, but true porcelain—fired at over 2300°F.—cannot. *Hard-paste* porcelain or *pâte dure* are two other names for true porcelain.

One of the most popular porcelain body formulas, fired in the stoneware range of C/9–10, is composed of 25 percent each of ball clay, feldspar, china clay, and silica. The following formulas are variations to experiment with.

PORCELAIN CLAY BODY FORMULAS

CB 1146 Casting Porcelain

Potash feldspar	40.0	Temperature	C/8–9
China clay	35.0	Shrinkage @ C/8	11.0%
Flint	20.0	Plastic scale	0
Calcium carbonate	5.0	Color	Creamy white (6)
	100.0	Absorption	4.4%
Soda ash	0.2		
Sodium bicarbonate	0.2		

CB 1147 Bone Casting Porcelain

Cornwall stone	40.0	Temperature	C/8–10
Bone ash	28.0	Shrinkage @ C/8	12.0%
E.P. kaolin	25.0	Plastic scale	0
Flint	7.0	Color	Bone white (32)
	100.0	Absorption	2.8%
Sodium silicate	0.1		

CB 1148 Mills Porcelain

E.P. kaolin	33.7	Temperature	C/9
Flint	31.6	Shrinkage @ C/9	15.0%
Kingman feldspar	23.2	Plastic scale	8.0
Kentucky ball #4	9.5	Color	Very pale gray (7)
Bentonite	2.0	Absorption	1.4%
	100.0		

CB 1149 Plastic Porcelain

Florida kaolin	25	Temperature	C/9–11
Ball clay	25	Shrinkage @ C/9	17.0%
Kingman feldspar	25	Plastic scale	8.5
Georgia kaolin	12	Color	Grayish white (31)
Flint	10	Absorption	1.9%
Bentonite	3		
	100		

II

Refractories

Hard Brick, Insulation Brick, Castables, Crucibles, Kiln Repair Paste, Exterior Insulation Coating, and Kiln Shelves and Posts

Thousands of years ago, when early ceramists baked their wares in open fires, potters most likely learned to improve the efficiency of the fire by placing stones around it. They found that efficiency could be further increased by stacking additional stones to build a low wall. However, as the primitive potters soon realized, many types of stone cannot withstand the heat of direct flames and will crack or even shatter. Finally, our ancestors discovered that the need for stones could be eliminated and the efficiency of firing even further improved if they built the fire in a dug-out pit in which the surrounding earth formed the walls. Able to withstand the heat and direct flames, the dirt would provide a better thermal insulating material than rock. Thus began the development of *refractories,* or units for firing ceramics.

Even with today's advanced technology of electric pyrometers, program-controlled firing, space age insulation, and improved knowledge of combustion, there is still a certain excitement in firing a kiln. Building one's own kiln from scratch is stimulating and personally rewarding. One can even take pride in making the brick for the kiln. The use of local clay materials and inexpensive equipment keeps the cost of building a kiln reasonable.

Several elements, such as oxides, carbonates, or nitrate forms, have high refractoriness and can be used for making brick. However, because of scarce availability and high cost, only silica, alumina, zircon, calcium, and magnesium are practical. Pure oxides or even carbonates are not found in sufficient amounts in nature for commercial extraction from the earth. The cheapest, most readily available refractory is clay. Clay, or kaolinite with the formula of $Al_2O_3 \cdot 2\ SiO_2 \cdot 2\ H_2O$, is not found technically (100 percent) pure. Traces of iron and alkalies are found in natural kaolin, and they lower its melting point to below 1780°c. Bricks of early kilns were made out of earthenware or red burning clay and worked fine for the lower temperatures. Many natural buff to brown earthenware clays are used for common house brick, and fire clays are used for insulating and fire brick. Other bricks are blends of different clays and minerals. The composition of the clay will determine the quality and temperature limitations of the brick.

The making of one's own refractories has its limitations in that it takes time, some equipment, and materials. The cost advantages are great, however, especially if one is going to build an unusually shaped or specialty kiln. For example, commercially manufactured rotary, cupola, and circle brick, made for round kilns, are very expensive, but their homemade equivalents are fairly inexpensive. The methods used in industry to make brick and other refractories can also be used in the studio.

Brick-Making Methods

In the *stiff mud* process, clay materials are mixed with water into a stiff consistency and put into a pug mill. The clay is then extruded through a die under tremendous pressure into a solid de-aired column of stiff semiplastic clay. Next, the column is wire-cut to the precise length desired. The individual potter can apply the same technique of mixing clay and extruding it through a pug mill or clay extruder.

A die with an opening slightly larger than the desired finished size is used, to allow for shrinkage during drying and firing. As the column is extruded, the clay is cut to length and placed on wareboards to dry. The *dry press* method is the use of slightly dampened clay that is placed into a mold and compressed under heavy pressure—approximately 2,500 pounds per square inch. A vacuum is attached to the mold to de-air the clay. The bricks produced by this method are extremely dense and uniform in size and shape. This is the most difficult process for individual potters to use, unless they have access to a press. *Press mold,* which is a slow process, is used primarily for special or custom-shaped bricks. Here the clay is first wedged and then placed into a wooden mold. The mold is then vibrated and bumped until the corners and bottom are filled with clay. A wooden hammer is used to compress and level the clay surface. Excess clay is removed from the surface by drawing a board across the top of the mold. The mold is then opened and the clay brick is set aside to dry. Potters can make arch brick, shelves, posts, special shapes, burner blocks, circular kiln brick, and other such refractories by constructing their own molds (an example of which is shown in Drawing 3). The industrial *air hammer* process is a variation of the press mold process in which an air hammer is used to tamp the relatively dry mixture into the molds. If the clay mixture is kept as dry as possible, the brick or other forms will be truer to dimensions and freer from warpage. The *casting brick* method is the least practical of the processes. The clay compound is mixed into a slurry and poured into a plaster mold. The mold is set out to dry, and after the clay stiffens a bit the mold is turned over and set up on ½-inch-high boards. When the clay has dried sufficiently, the form will drop from the mold onto the board. (Triangular porcelain kiln posts are also made by this method. See pages 81–82.)

Regardless of the method used, all bricks are set out to air-dry. Some bricks will take a long time to dry. After drying, the bricks are stacked in a kiln to be fired; they are arranged in a criss-cross pattern in which the bricks are first placed across the hearth and then on top of each other, with ⅜- to ¾-inch air passages between the layers and between individual bricks. The criss-cross placement permits several layers of brick to be fired at once and facilitates greater evenness of temperature and air circulation during the firing.

Drawing 3
Wooden Frame for Making Bricks (Press Mold Method)

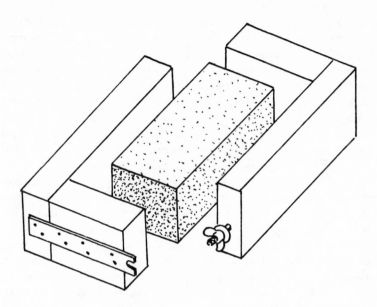

Both preheating and firing must be slow to allow the moisture and organic materials to escape gradually. The organic materials that make up the insulation brick burn slowly and will cause smoke and odors.

Hard Brick

Buying hard brick is generally more practical than making one's own. Used fire brick can be obtained from fireplaces of condemned buildings, from furnaces of boilers, from incinerators, and from steel and iron furnaces. Numerous other sources of used brick are available, and a check through the telephone-book yellow pages will reveal many of them. Used brick can be a savings, but there are many different types of hard brick, so if the refractoriness temperature of a particular brick is unknown, a sample brick should be test-fired in a kiln. For instance some white and yellow house brick, which would soften in firing, looks like fire clay brick and could be mistaken for such.

There are many types of hard brick. *Low-duty brick* is used for temperatures under 1500°F. *Medium-duty brick* is used for fireplaces, barbecue grills, moderate-duty boiler service, annealing furnaces, and back-up liners for heavy-duty furnaces. *High-duty brick* is the standard high-temperature brick used for furnaces and boilers. *Super-duty brick* (semisilica and flint fire clay) is used for blast furnaces, stoves, open hearths, soaking pits, and ceramic kilns. *High-alumina brick* has an alumina content ranging from 50 to 80 percent and is used in forge furnaces, in rotary cement and lime kilns, and in high-temperature kilns. *Extra-high-alumina brick,* containing from 85 to 99 percent alumina, is an extremely expensive specialty brick used in very high temperature applications of up to C/42. *Mullite brick* contains no excess silica and is used in glass furnaces, tunnel kilns, burner blocks, and incinerators. *Zircon brick,* containing a high percent of zircon, is used for acid slag tanks, platinum-melting crucibles, and frit-melting furnaces. *Chrome brick,* made of periclase and chrome-concentrated ore, is used in electric melting furnaces of brass, ferro, copper, and copper alloys. Specialty brick like *high-purity magnesia* is used to line the melting furnaces of corrosive alloys.

For most individually made brick a compound of grog, fire clay, kaolin, and ball clay, in various proportions, is recommended. Iron-free clays and grogs are desirable. Often brick made of 100 percent regional fire clays is the cheapest and is usually satisfactory. Local brick, earthenware, and sewer pipe clays, when mixed with fire clay, also make an inexpensive and serviceable brick.

The formulas listed are for grog/fire clay mixtures to which other minerals—such as dolomite, periclase, alumina hydrate (expensive), zircon, and silicon carbide—can be added to give the brick higher temperature refractoriness.

HARD-BRICK FORMULAS

R 101 Hard Brick

Ione-grain grog,		Temperature		C/10
30–80 mesh	80	Shrinkage @ C/10		14%
Ball clay	20	Plastic scale		0
	100			

R 102 Hard Brick

Grog, 60–80 mesh	70	Temperature		C/10
Plastic fire clay	20	Shrinkage @ C/10		7.0%
Ball clay	10	Plastic scale		0
	100			

R 103 Hard Brick

Grog, 60–80 mesh	60	Temperature		C/10
Plastic fire clay	30	Shrinkage @ C/10		10%
Ball clay	10	Plastic scale		0
	100			

R 104 Hard Brick

Grog, 60–80 mesh	50	Temperature		C/10
Plastic fire clay	40	Shrinkage @ C/10		13%
Ball clay	10	Plastic scale		3
	100			

R 105 Hard Brick

Grog, 60–80 mesh	40	Temperature		C/10
Plastic fire clay	40	Shrinkage @ C/10		14%
Ball clay	20	Plastic scale		3
	100			

Insulation Brick

Insulation brick, also known as *soft brick,* is used for both insulation and structural support, either alone or as a back-up for hard brick. One type of soft brick is made from *diatomaceous earth (Fuller's earth),* a natural material that is quarried and then cut to shape. Diatomaceous earth is formed from natural deposits of small sea diatoms, producing a light and porous silicon dioxide material. Although weak in structure and limited in temperature range, brick made from diatomaceous earth is well suited for use as back-up insulation. A second type of soft brick is made from a mixture of diatomaceous earth and clay. The third type of soft brick is a combination of kaolin and fire clays.

Two methods are used to give soft brick the light weight and porosity which are necessary insulation properties. In the chemical method, air bubbles are introduced into the wet brick mixture by chemical action to produce a uniform-size, multicellular, lightweight structure with high strength and low density. The resulting bricks are like compressed hollow grains of clay. In the more commonly applied technique, the brick is made from a mixture of screened sawdust (or other organic material that has few impurities) and clays. Coke, coal, ash, straw, and nutshells are some of the materials that can be used. During firing, the organic material burns away, leaving small air pockets in the brick.* The air pockets make excellent insulators that prevent thermal conductivity (heat loss). The lightweight quality of the brick depends on the amount of organic material added to the mixture. The more organic material used, the less dense the brick, and vice versa.

The formulas listed on the following pages are examples of possible types of insulating brick. Screened walnut shells were the primary organic material used in these formulas. Crushed and screened nutshells are available from several sources and have the advantage of being ready to use, although they are slightly expensive. Screened sawdust is also an excellent material and is usually either free or very inexpensive. Many local lumber yards will gladly give it away

---

* It is important that the organic materials be screened so that uniform air pockets will be produced and uniform temperatures can be maintained, preventing hot spots.

just to be rid of it. Because walnut shells are heavier than sawdust or straw, formulas given in this book should be adjusted if the latter two materials are used in place of walnut shells. When the formulas are prepared, the clays and other ingredients should be mixed before the organic materials are added. The complete mixture should be allowed to age a day or so, and the brick then made by extruding or press-molding.

INSULATION BRICK FORMULAS

R 106 Insulation Brick

Walnut shells	60.0	Temperature	C/10
Plastic fire clay	30.0	Shrinkage @ C/10	15%
Ball clay	10.0	Plastic scale	0
Sodium silicate	0.5	Structure	Soft
	100.5		

R 107 Insulation Brick

Walnut shells	60	Temperature	C/10
Lincoln fire clay	35	Shrinkage @ C/10	16%
Tennessee ball	5	Plastic scale	0
	100	Structure	Soft

R 108 Insulation Brick

Walnut shells	50	Temperature	C/10
Lincoln fire clay	45	Shrinkage @ C/10	14%
Tennessee ball	5	Plastic scale	0
	100	Structure	Soft

R 109 Insulation Brick

Walnut shells	50	Temperature	C/10
E.P. kaolin	40	Shrinkage @ C/10	17%
Tennessee ball	10	Plastic scale	0
	100	Structure	Soft

R 110 Insulation Brick

Walnut shells	50.0	Temperature	C/10
Plastic fire clay	40.0	Shrinkage @ C/10	14%
Ball clay	10.0	Plastic scale	0
Sodium silicate	0.5	Structure	Soft
	100.5		

R 111 Insulation Brick

Walnut shells	50.0	Temperature	C/10
Plastic fire clay	35.0	Shrinkage @ C/10	15%
Ball clay	15.0	Plastic scale	0
Sodium silicate	0.5	Structure	Soft
	100.5		

R 112 Insulation Brick

Walnut shells	45	Temperature	C/10
Plastic fire clay	35	Shrinkage @ C/10	16%
Ball clay	15	Plastic scale	0
Talc	5	Structure	Soft
	100		

R 113 Insulation Brick

Walnut shells	40	Temperature	C/10
Plastic fire clay	40	Shrinkage @ C/10	15%
Ball clay	15	Plastic scale	0
Talc	5	Structure	Soft
	100		

Castable Refractories

Monolithic kilns have greater combustion efficiency, longer life, and more trouble-free surface area than hard-brick kilns. In typical brick construction, the brick joints erode and chip, causing flakes to fall into the kiln. Also, the life of the brick is shortened by the erosion created by the exposure of a greater surface area to flames and their gases. Parts of the kiln, such as the door, the arch, and the firebox, may be cast; or the entire structure may be cast. Another possibility is to build a conventional hard-brick kiln and cover the inside with an insulating castable, which should be mixed to trowel consistency and applied in thicknesses ranging from 2 to 6 inches. (Kilns used outside that do not have a roof can be given a waterproof coating of portland cement, fire clay, and silica sand. Wire mesh is used to hold this mixture to the outside of the kiln. (See pages 79–80.)

In casting the kiln, entirely or in parts, a wooden frame (mold) coated with heavy oil or wax is used; the technique is the same as in concrete construction. The casting mixture is poured into the frame. After the mixture has set, the frame is removed (and used for additional castings if desired). This procedure is followed for each kiln part that is to be cast. Once the kiln is completed, it will take several days for the moisture to evaporate. When dry, the kiln is fired *very* slowly to remove the water and organic materials and to set the castable.

Some of the formulas listed use screened walnut shells, but screened vermiculite, sawdust, or straw can be substituted. Before the building of a large kiln or other structure is undertaken, a *raku* or test kiln should be made to test both the construction techniques and the castable to be used.

CASTABLE-REFRACTORY FORMULAS

R 114 Castable Brick

Lincoln fire clay	60	Temperature	C/9
Walnut shells	20	Shrinkage @ C/9	8.0%
Mullite	10	Plastic scale	2
Portland cement	10	Strength	Sturdy
	100		

R 115 Castable Brick

Walnut shells	40	Temperature	C/9
Coarse grog, 10 mesh	20	Shrinkage @ C/9	6.0%
Lincoln fire clay	20	Plastic scale	0
Portland cement	10	Strength	Weak
Calcined kaolin	10		
	100		

R 116 Castable Brick

Mullite	40	Temperature	C/9
Coarse grog	20	Shrinkage @ C/9	2.0%
Walnut shells	20	Plastic scale	0
Tennessee ball	10	Strength	Soft
Portland cement	10		
	100		

R 117 Castable Brick

Walnut shells	30	Temperature	C/9
Lincoln fire clay	30	Shrinkage @ C/9	5.0%
Mullite	20	Plastic scale	0
Ball clay	10	Strength	Medium-hard
Portland cement	10		
	100		

R 118 Castable Brick

Mullite	30	Temperature	C/9
Coarse grog, 10 mesh	25	Shrinkage @ C/9	1.0%
Walnut shells	15	Plastic scale	0
Ball clay	15	Strength	Weak
Portland cement	15		
	100		

R 119 Castable Brick

Mullite	27	Temperature	C/9
Walnut shells	27	Shrinkage @ C/9	2.0%
Coarse grog	18	Plastic scale	0
Ball clay	14	Strength	Medium-hard
Portland cement	14		
	100		

R 120 Castable Brick

Pea grog	25.0	Temperature	C/10
Fire clay	25.0	Shrinkage @ C/9	4.0%
Vermiculite	18.8	Plastic scale	0
Portland cement	12.5	Strength	Medium
Sawdust	12.5		
Alumina hydrate	6.2		
	100.0		

Crucibles

Crucibles are highly refractory vessels used in ceramics for melting stains, glass, frits, and enamels. (Jewelry makers, sculptors, and other craft workers use crucibles to melt lead, silver, gold, bronze, or other metals.) A very stiff, highly grogged clay compound is used to make crucibles, whether hand-built, thrown, cast, or (as most are) press-molded. The wall thickness must be uniform and without any defects, such as hairline cracks, bubbles, thin spots, and holes. The crucible is slowly dried and then fired at high temperature.

To use the crucible, the potter dry-mixes glass, frit, stain, or enamel ingredients and then places them in the crucible to be melted. The crucible may then be heated directly with a torch or placed in a test, enameling, or firing kiln. When the maturing temperature of the substance being melted is reached, and all the bubbling has stopped, the crucible is removed with tongs from the kiln, and the hot liquid is poured into cold water. The glass-like liquid mixture will shatter into small fragments which are then placed in a ball mill and ground to 60-mesh (or finer) grains. To reduce thermal stress and prolong its life, the hot crucible is returned to the warm kiln to cool slowly as the kiln cools.

Most crucible formulas are mixtures of kaolin, fire clay, and grog. The formulas given here are variations of this combination.

CRUCIBLE BODY FORMULAS

R 121 Crucible Body

Georgia kaolin	40	Temperature	C/9
Calcined kaolin	20	Shrinkage @ C/9	14%
Tennessee ball	20	Strength	Hard
Fine grog	10	Structure	Dense
Flint	10		
	100		

R 122 Crucible Body

Plastic fire clay	42	Temperature	C/6–9
Fine grog	20	Shrinkage @ C/6	13%
Sand	15	Strength	Hard
Flint	15	Structure	Dense
Bentonite	8		
	100		

R 123 Crucible Body

Plastic fire clay	50	Temperature	C/9
Calcined kaolin	30	Shrinkage @ C/9	15%
Fine grog	20	Strength	Hard
	100	Structure	Dense

R 124 Crucible Body

Plastic fire clay	50	Temperature	C/9
Fine grog	30	Shrinkage @ C/9	14%
Calcined kaolin	20	Strength	Hard
	100	Structure	Dense

R 125 Crucible Body

Plastic fire clay	50	Temperature	C/9
Fine grog	30	Shrinkage @ C/9	14.5%
Talc	20	Strength	Hard
	100	Structure	Dense

R 126 Crucible Body

Plastic fire clay	50	Temperature	C/6–9
Calcined kaolin	20	Shrinkage @ C/9	14%
Fine grog	20	Strength	Hard
Talc	10	Structure	Dense
	100		

R 127 Crucible Body

Plastic kaolin	50	Temperature	C/9
Talc	20	Shrinkage @ C/9	15%
Fine grog	20	Strength	Hard
Spodumene	10	Structure	Dense
	100		

Kiln Repair Paste

At one time or another, every kiln will need minor or major repairs. The type of repair paste used will depend upon the kind of kiln repair being made. There are several types of repair pastes, each of which has certain qualities that make it best suited to a particular type of kiln repair.

To mend holes or broken-out corners, a piece of insulation brick is crushed and then mixed with moist repair paste. The hole is first painted with the moist paste, and then the brick–paste mixture is packed into the hole or damaged area. Repairs made on large areas of the kiln or on badly damaged brick require brick replacement, as does restoration of missing brick. The replacement brick must be cut to fit the affected area and a thin coating of paste applied to both the replacement brick and to the area into which the brick is to be fitted. The third type of kiln repairs—those made on cracks— are the most difficult. In repairing cracks, a thin mixture of paste is made and, with a fettling or other thin-bladed instrument, is forced as deep as possible into the crack. Applying several thin coatings to fill up kiln holes and cracks is better than applying one thick coating.

The following paste formulas should be sufficient for both minor and major kiln repairs.

KILN REPAIR PASTE FORMULAS

R 128 Kiln Repair Paste

Lincoln fire clay	45	Temperature	C/9
Calcined kaolin	20	Shrinkage @ C/9	Little shrinkage
Talc	14		with thin coating
Fine grog	9		High shrinkage with
Walnut shells	9		thick coating
Sodium silicate	3		
	100		

R 129 Kiln Repair Paste

Lincoln fire clay	30	Temperature	C/9
Fine grog	20	Shrinkage	Same as for R 128
Walnut shells	20	Strength	Hard
Calcined kaolin	17	Note	Apply thin coats
Talc	10		to build thickness
Sodium silicate	3		
	100		

R 130 Kiln Repair Paste

Tennessee ball	25	Temperature	C/9
China clay	25	Shrinkage	Same as for R 128
Flint	25	Strength	Hard and dense
Oxford feldspar	25	Note	Apply thin
	100		

Exterior Insulation Coating for Kilns

A kiln that is kept outside without any type of cover will require a coating to protect it from the elements. Common house brick is used for industrial kilns. Smaller kilns, like cantinary arch, are cov-

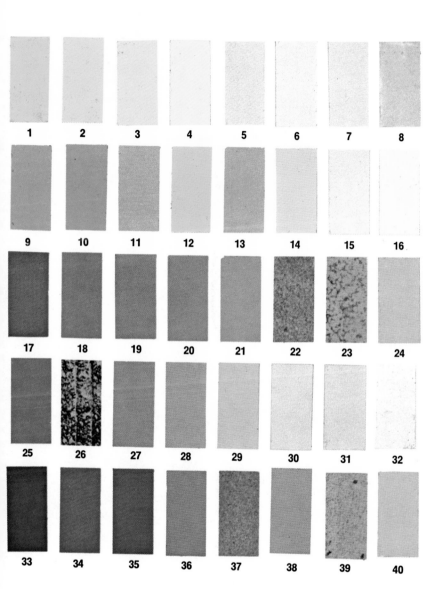

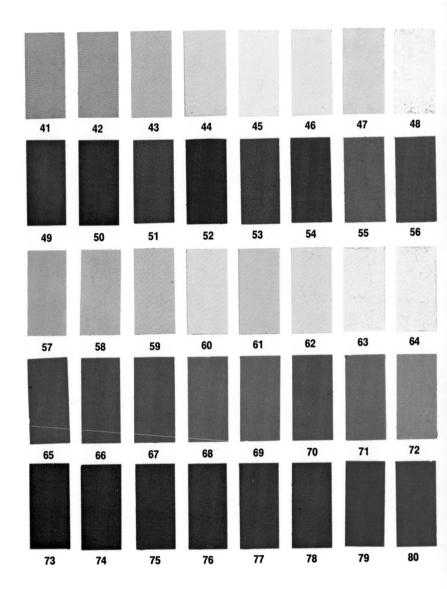

41 42 43 44 45 46 47 48

49 50 51 52 53 54 55 56

57 58 59 60 61 62 63 64

65 66 67 68 69 70 71 72

73 74 75 76 77 78 79 80

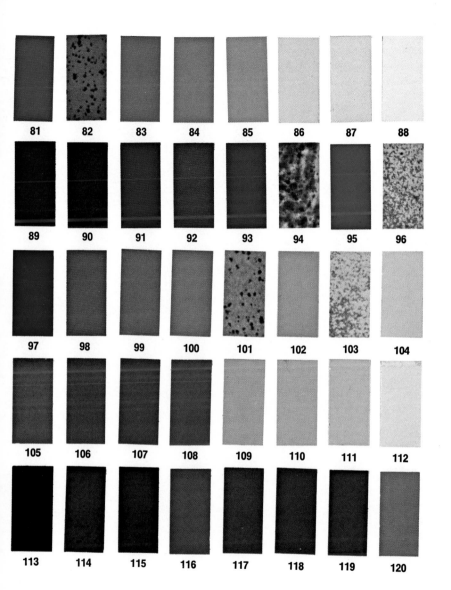

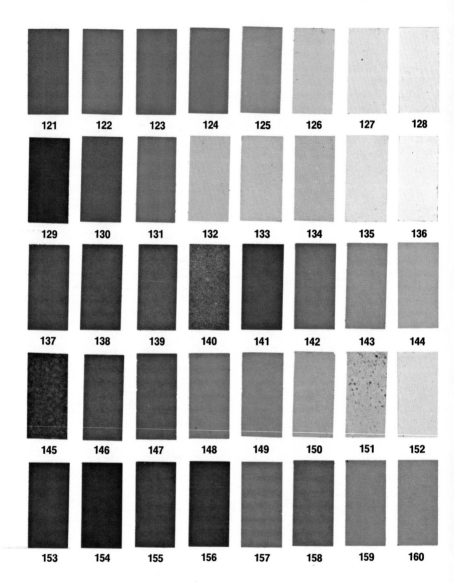

121 122 123 124 125 126 127 128

129 130 131 132 133 134 135 136

137 138 139 140 141 142 143 144

145 146 147 148 149 150 151 152

153 154 155 156 157 158 159 160

ered with a portland cement–based coating to provide both insulation and waterproofing. To secure the coating, chicken or expanded wire mesh is attached to the kiln, and a mortar of a thick consistency is applied to the kiln by forcing it through the wire mesh. A second coat, again a mortar of thick consistency, is applied both to cover the wire mesh and to increase the kiln-wall thickness to between ¾ inch and 3 inches (as desired). The third coating is of a slightly thinner consistency and does not contain walnut shell, sawdust, straw, or vermiculite. It is applied over the kiln surface to a ¼-inch-thickness to provide a finish coat that is both smooth and waterproof.

The formulas listed contain walnut shell, but vermiculite, sawdust, or straw can be substituted. A weatherproof-only coating could be prepared from any of the formulas listed, minus the shells (vermiculite, sawdust, or straw); or formula R 134 could be used.

EXTERIOR INSULATION COATING FORMULAS

R 131 Exterior Insulation Coating

Portland cement	60	Shrinkage	Little (under 2%)
Walnut shells	20	Strength	Moderate
Mullite	10		
Coarse grog	10		
	100		

R 132 Exterior Insulation Coating

Portland cement	50	Shrinkage	Little (under 2%)
Walnut shells	20	Strength	Moderate
Coarse grog	20		
Mullite	10		
	100		

R 133 Exterior Insulation Coating

Portland cement	40	Shrinkage	Little (under 2%)
Walnut shells	30	Strength	Moderate
Coarse grog	20		
Mullite	10		
	100		

R 134 Exterior Weatherproof Coating

Portland cement	40	Shrinkage	Little (under 2%)
Silica sand	30	Strength	Dense and hard
Mullite	30		
	100		

Kiln Shelves and Post Bodies

Most shelves used in ceramics are made either of mullite or of silicon carbide compositions. Silicon carbide shelves are strong and have little warpage, but they are expensive and extremely difficult for studio potters to make because their construction, like that of high-temperature mullite shelves, is an exacting process which requires heavy presses for molding the shelves and precise drying and firing cycles. Shelves for *raku* and bisque-firing kilns, as well as small shelves for high-temperature firing, on the other hand, can be easily constructed in the studio.

The easiest method of making shelves is to use a press mold and semimoist clay. Heavy-duty plaster press molds are thicker than ordinary casting press molds and have reinforcement rods or chicken wire (Drawing 4). In making a mold, a kiln-shelf pattern—made 5

Drawing 4
Top and Cross-Section Views of Kiln-Shelf Plaster Press Mold

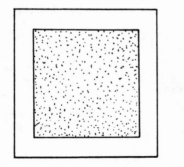
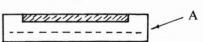

A. Reinforcement rod or chicken wire

to 10 percent larger to allow for shrinkage of the kiln shelf mixture—is built of the clay and placed on a flat, nonporous surface; a wooden frame is constructed around it, with a 2-inch clearance allowed for around and above the shelf. Plaster is then poured into the clearance spaces, and when it reaches a soft-ice-cream consistency, chicken-wire mesh is positioned in the plaster. After the plaster has set, the wooden framing is removed and the clay shelf dug out. The mold is then set aside to cure.

To use the mold, a kiln-shelf clay mixture is made, with just enough water to allow it to be wedged and rolled out. A slab of the clay is placed in the mold and hammered with a wooden mallet to compress it and force it into shape. Next, a board is drawn across the mold, scraping away excess clay. Recessed areas are then filled and the surface is sponged smooth. The clay will shrink enough within two hours to enable it to be removed from the mold. When the clay has shrunk sufficiently, a piece of cloth and a wareboard are placed over the mold. The mold is then turned over onto the wareboard and lifted, leaving the clay shelf. The cloth will prevent the clay from sticking to the wareboard. The slab is set aside to dry slowly and is turned over several times during the drying to prevent warpage. When completely dry, it is high-fired.

Individually homemade shelves are to be made thicker than commercial shelves. For example, a 14-inch-square shelf for bisque and *raku* firing would be ¾ to 1 inch thick, and for high temperatures 1 to 1¼ inches thick. The shelves are turned over after several firings to correct bending.

Posts are made commercially by casting or extruded techniques; the same techniques can be applied in the studio. For casting a plaster mold, one-part or two-part molds are used. Possible mold shapes are shown in Drawing 5. Molds are made 14 inches long to allow for shrinkage, and the leather-hard posts may later be cut to the various lengths desired. The formulas for slipcasting posts are similar to those for porcelain. In fact, C/12 casting porcelain can be used at earthenware and stoneware temperatures. Slip is poured into the mold and permitted to build up to a thickness of ⅜ inch; the excess is poured out. If the cast is too thin, the post will warp or sag in firing and thus become unusable. When the clay has set, it is removed from the mold, placed on a cloth-covered wareboard, and cut to desired lengths of 1, 2, 3, 4, 5, 6, 8, 10, or 12 inches.

Extruded posts are often stronger and longer-lasting than cast posts. A metal templet is made for the post (see Drawing 5). Most hand extruders can be used to make hollow tubes. The clay is mixed, thoroughly wedged, and extruded. Lengths of extruded posts are placed on cloth-covered wareboards to stiffen slightly; then they are cut to desired lengths. A board or other straight edge is used to make sure that the post is straight and tops and bottoms are cut true. Periodically during the drying process, the posts are turned to permit even drying. When completely dry, they are high-fired.

Drawing 5
Cross-Section View of Shapes of Kiln Posts for Both Casting and Extruded Techniques

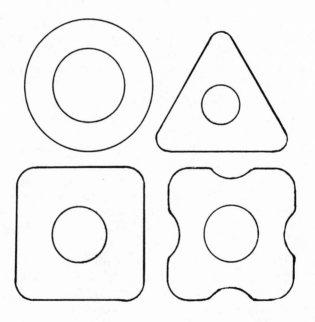

KILN-SHELF AND POST BODY FORMULAS

R 135 Shelves/Posts

Mullite	65	Temperature	C/11
Calcined kaolin	10	Shrinkage @ C/11	3.0%
Ball clay	10	Plastic scale	0
Fine grog	10	Color	Light tan (30)
Talc	5	Absorption	16%
	100		

R 136 Shelves/Posts

Mullite	60	Temperature	C/11
Calcined kaolin	20	Shrinkage @ C/11	4.0%
Ball clay	10	Plastic scale	0
Talc	10	Color	Sand tan (30)
	100	Absorption	16%

R 137 Shelves/Posts

Mullite	50	Temperature	C/11
Calcined kaolin	20	Shrinkage @ C/11	2.0%
Talc	20	Plastic scale	2.0
Fine grog	10	Color	Light tan (29)
	100	Absorption	18%

III

Raw Stains

Stains

The term *stain* in ceramics is a broadly used expression that applies to certain ceramic colorants. With some exceptions, stains are composed of metallic oxides and carbonates, spinels, or fritted ceramic pigments, or ceramic metallic oxide mixtures; they are used in coloring engobes, clay bodies, glazes, ceramic decals, clay surfaces, underglazes, screen printing inks, and overglazes. The various types of stain mixtures include those that contain cobalt, copper, chrome, manganese, iron, nickel, or uranium oxides and carbonates, used alone or in different combinations.

Metallic oxides and carbonates are the main colorants used in ceramics. Many ceramists use the amount of metal in a formula to determine whether an oxide or a carbonate is to be used. If a formula calls for less than 2 percent of a metal—such as copper, for example—then the carbonate is used. If there is more than 2 percent metal, then the oxide is used. The carbonate is composed of finer particles and will disperse more readily than the more coarsely grained oxide. Copper oxide, unless ground, will often cause specks in the glaze. This is particularly true for copper reds, where only 0.5 to 1 percent is used.

Spinel is a natural mineral found in small deposits. It was created by a reaction between MgO and Al_2O_3 (magnesium aluminate), and between BeO and Al_2O_3 (beryllium aluminate), while these minerals were forming. Artificial spinel is the composition $RO \cdot R_2O_3$ (alkaline oxide and an amphoteric oxide that acts as an acid). Cobalt, chrome, zinc, nickel, iron manganese, and cadmium make up the RO group, while alumina, chrome, and iron constitute the R_2O_3 group. Synthetic spinels are made by grinding together the various oxides that make up the two groups with a small amount of boric acid, which helps to accelerate the reaction. The ground mixture is then calcined for several hours at 1800° to 2300°F, depending upon the compound. Afterwards it is crushed and wet-ground, with a small amount of acid added to aid the grinding. Several washings are necessary to remove the acids and soluble salts.

Fritted ceramic pigments are spinel-like metallic combinations. The ceramic materials and oxides are ground, calcined, reground, and washed, just like spinels. Fritted ceramic pigments are known as *commercial stains;* examples are chrome-tin pink, antimony yellow, zirconium-vanadium blue, tin-vanadium yellow, chrome green, and zirconium-vanadium-indium orange.

The minerals alumina, zinc, tin, antimony, zirconium, vanadium, and/or zirconium are used with *ceramic metallic oxides.* Examples are cobalt and alumina for blue, copper and alumina for green, chrome and zinc for gray-green, iron and zinc for brown, chrome and tin for pink, and uranium and alumina for orange.

Raw stains are mixtures of metallic oxides, carbonates, ceramic materials, and/or spinel or fritted ceramic pigment in the raw state—that is, the material is not fritted. Instead, the oxides and other ceramic materials are thoroughly blended to form a ready-to-be-used mixture. Even some of the commercial stains are raw, particularly the ceramic blacks.

Spinel, fritted ceramic pigments, metallic oxides, and raw stains are all employed in the same way, either alone as a colorant or mixed with glazes and clays. By themselves they are applied to the surface of raw clay, dry clay, bisqueware, or a glaze. A small amount of glue will help the stain adhere to the clay body. Raw stains have several particular advantages: they are quickly and easily made; they are not as expensive as commercial stains; several ounces—or even several pounds—of a favorite stain can be prepared in advance and stored; and they offer a wider range of colors than is possible with plain metal oxides.

Making Raw Stains

The only equipment necessary for making raw stains is an accurate scale and a mortar and pestle. The materials are weighed—usually in amounts to total 500 grams (slightly over 1 pound) to 1500 grams—mixed thoroughly, sifted several times through a 120-mesh screen, and then stored in a secure container with a label. All ceramic materials should be labeled clearly, for the loss of even a few ounces of some materials (such as cobalt) can be very expensive. It is a common occurrence, in colleges and studios, for several pounds of favorite raw stains to be made up at a time. Not only is such a practice convenient and efficient, but it provides greater control over color—the ceramist can determine the colorant proportions more readily—and expedites glaze making.

Influences on Colorants

1. High temperatures (which cause some compounds to decompose)
2. Acid, neutral, or base condition of the glaze compound
3. Degree of activity of the glaze flux
4. Reductive, neutral, or oxidative character of firing atmosphere
5. Location in kiln (hot spots, cold spots, path of flame)
6. Presence of tin, chrome, or cobalt in adjacent pots (Vapors will give pink tints.)
7. Time of day of kiln firing (During heavy fuel-use periods, gas pressure drops slightly, necessitating a longer firing time.)
8. Length of soaking time of firing (permits the materials in the glaze to fuse completely)
9. Glaze thickness (A certain percent of the glaze and stain will react. A thin glaze coating is particularly vulnerable.)
10. Composition of clay body (Iron or other impurities will

influence the color. In some cases, as with celadon, this is desirable; but it is disastrous on red, pink, and yellow colors.)

11. Mood of the kiln gods (very important consideration)

Raw Stains at High Temperatures

Ceramic raw-stain colorants consist of solid particles, with each particle being made up of metallic oxides and stabilizing materials. Such colorants are relatively inert and have few differences in application; however, some are affected by high temperatures and, in some cases, absorption of the colorant by a particular glaze or clay body will cause decomposition, burning out, or loss of color identity.

Obtaining successful results with certain volatile (unstable) stains—in particular, bright and clean red, orange, yellow, pink, and some pastel colors—requires the selection of suitable glaze, firing temperature, and application method. Patience and accurate notetaking are essential to obtain reproducible and consistent colors for each firing. Some colorants, like chrome and cobalt, are very stable in most glazes at high temperatures; but other colorants, like copper and iron, will readily assimilate or release oxygen molecules. To achieve stable colors at C/4 or higher temperatures, one must consider the following:

Most red, orange, and yellow commercial or raw stains are used in amounts that vary from 5 to 17 percent. A higher percent of stain in the glaze will make it more resistant to heat, and, to compensate, the glaze will need to be more fluid. A white engobe should first be applied over iron or dark-bodied clay to maintain the pastel or bright colors. Applying the stain on the clay surface as an underglaze with a clear glaze over it, rather than putting the stain in the glaze, will produce a deeper color, particularly with the chrome-green hues. Also, some stains if mixed in the glaze will render it opaque or flat.

Metallic Oxides in Raw Stains

Raw stains are mixtures of metallic oxides and carbonates, sometimes with a stabilizer (or carrier) added. Metallic oxides are the coloring agents in stains; each oxide has its own color, stability, and intensity. The following is a list of the most frequently used oxides. The percentages given for each oxide are a guide for general use.

Antimoniate of lead, $Pb_2(SbO_4)_2$. Not often used. When fritted, it produces Naples Yellow, with a limited temperature range of up to 1920°F. 4 to 12 percent.

Antimony oxide, Sb_2O_3. The ore is mined in China, Mexico, and Bolivia. It functions as an opacifier and coloring agent to give weak whites. Yellow colors are possible if it is used in lead glazes. Unstable above C/1. 5 to 18 percent. TOXIC.

Cadmium, Cd (usually cadmium sulphide, CdS) and *Selenium,* Se. Both are the basis of low-temperature reds, oranges, and yellows in leadless glazes. It is fugitive over 1950°F. 3 to 10 percent. TOXIC.

Ceramic black (mixture of various oxides). A raw stain made of various oxides, with black iron, manganese dioxide, and copper dominating. 8 to 20 percent.

Chrome oxide (green) Cr_2O_3. The ore is mined in South Africa, Turkey, Rhodesia, and Russia. It is opaque in amounts above 1 percent. In low-temperature lead glazes, it produces yellows, oranges, and reds; in zinc glazes, brown; in low- to mid-temperature glazes with tin, pink; with cobalt, blue-green; and in most other glazes, typical chrome green colors. 0.5 to 3 percent.

Cobalt carbonate, $CoCO_3$; oxide, CoO. The most powerful colorant. As little as 0.2 percent will produce a blue color. The various shades of blue it produces are stable colors. If manganese is added, purples are produced. 0.2 to 2.5 percent.

Copper carbonate, $CuCO_3$; oxide, CuO; cuprous (red), Cu_2O. The oldest colorant used. In lead glazes, it produces grass greens; in alkaline glazes, turquoise (Egyptian paste) and blues; and in reduced glazes, copper reds (rouge flambé, sang-de-boeuf). In amounts above 2 percent it is an active flux. 0.5 to 7 percent.

Ilmenite (ferrous titanate), $FeO \cdot TiO_2$. This mineral is obtained from the beach sands of Australia. An impure (iron-bearing) form of titanium is used, in granular form, for dark spots and specks. Its powder form is used like rutile. 0.1 to 5 percent.

Iron oxide (black), Fe_3O_4; (red), Fe_2O_3. In oxidation, it produces warm cream, brick red, yellow, brown, and black; in reduction, the colors become cool and the blue, gray, and green of celadon; and in saturated amounts of 8 to 20 percent, depending upon the glaze and temperature, goldstone, tiger's-eye (aventurine), and tenmoku (oil spot, hare's fur, luster black, and iron spot). 1 to 20 percent.

Iron chromate, $Fe_2O_3 \cdot Cr_2O_3$. Used primarily for producing gray and black in glazes and clay bodies. 3 to 10 percent.

Lead chromate, $PbO \cdot PbCrO_4$. Used in low-temperature glazes and stains for deep greens and Persian and coral reds. 1 to 5 percent.

Manganese dioxide, MnO_2; carbonate, $MnCO_3$. In lead glazes, it produces browns; in alkaline glazes, purples; with cobalt, violet; and with iron, chrome, copper, and cobalt, it produces clean blacks. The carbonate is a strong flux. 2 to 8 percent.

Nickel oxide, NiO; carbonate (green), $NiCO_3$. The predominant color is gray, but yellow, brown, and blue are possible. It is often used to tone down cobalt and copper colors. 0.5 to 3 percent.

Potassium bichromate, $K_2Cr_2O_7$. Used for low-temperature lead glazes producing oranges and reds. Temperatures up to 1650°F. 1 to 10 percent.

Raw sienna. A natural mineral with high iron content used for tan, buff, and brown colors. 1 to 12 percent.

Rutile, TiO_2. An impure form of titanium containing small amounts of iron. Cream, buff, tan, brown, and blue (above 6 percent) colors are produced. In amounts of more than 4 percent the color will be variegated in most glazes. 1 to 15 percent.

Tin oxide, SnO_2. Stable and consistent opacifier. With 0.5 percent copper carbonate, it produces copper red; and at over 10 percent, it takes on the appearance of coagulated white glaze. To make clean whites, use 3 to 5 percent for semiopaque and 5 to 10 percent for opaque. 2 to 10 percent.

Titanium oxide, TiO_2. Used as an opacifier, as a matting agent, and as a source for white. In amounts over 5 percent, it is used for matting and opacifying; and at 5 to 15 percent, for crystalline matt (satin matt) or crystal glaze. 2 to 15 percent.

Umber (raw and burnt). A naturally occurring hydrated iron oxide used for rust and red-brown colors in clay bodies and engobes. 3 to 15 percent.

Uranium oxide, U_3O_8. Used at temperatures up to 1850°F for reds, oranges, and yellows in lead and/or boron-based glazes. In salt and other glazes, it will give a weak yellow-orange at higher temperatures, even to C/10. 2 to 12 percent. TOXIC.

Vanadium pentoxide, V_2O_5. Seldom used by itself or in the raw state. Calcined as a stain with a tin mixture (vanadium-tin yellow), by itself (vanadium yellow), or zirconium (vanadium-zirconium blue or turquoise). In reduction for blue-blacks. 3 to 10 percent.

Zinc oxide, ZnO. This is a crystal former and an opacifier; it produces warm whites and at high temperatures is a flux. Saturated zinc of over 10 percent will produce crystalline matt (smooth matt) or crystals. 5 to 25 percent.

Zirconium oxide, ZrO_2 (zircopax, $ZrSiO_4$; zircon spinel, $ZrO_2 \cdot SiO_2 \cdot Al_2O_3 \cdot ZnO$. A fractory material used for making kiln furniture. Used as an opacifying agent for brilliant colors and whites. Used in its natural form or fritted with other materials. 5 to 18 percent.

RAW-STAIN FORMULAS

Each of the raw-stain formulas listed below was mixed in 100-gram units and dry-ground; sufficient water was added to bring the mixture to painting consistency, and then the stain was painted on three different white clay test bars. A glaze was painted over one-half of the stain on each bar. The three glazes used were C/015 high lead (glaze A), C/04 colemanite base (glaze B), and C/8–10 cornwall stone base (glaze C). The formulas are as follows:

Glaze A C/015

White lead	65
Calcined borax	20
Kaolin	15
	100

Glaze B C/04

Colemanite	38
Kingman feldspar	35
Barium carbonate	15
Flint	12
	100

Glaze C C/8–10

Cornwall stone	80
Whiting	14
Nepheline syenite	6
	100

The test bars were then fired to the appropriate temperature. When they had cooled, the color of both sides of each test bar (glazed and unglazed) was recorded. The resulting colors of the formula, each fired at a different temperature, are listed below.

S 101 White Clay Body

Tin oxide	94			
Antimony oxide	6	A	Color @ C/015	White (32)
	100		Color (glaze over)	Goldenrod (13)
		B	Color @ C/04	White (32)
			Color (glaze over)	White (7)
		C	Color @ C/8–10	White (32)
			Color (glaze over)	White (7)

S 102 Green

Tin oxide	90	Color @ C/015	Light chrome green (133)
Chrome oxide	10		
	100	Color (glaze over)	Mottled red, yellow, and green
		Color @ C/04	Light green (133)
		Color (glaze over)	Olive green (142)
		Color @ C/8–10	Light chrome green (133)
		Color (glaze over)	Gray-green (122)

S 103 Blue

Tin oxide	88.0	Color @ C/015	White (7)
Boric acid	8.5	Color (glaze over)	Mottled greens/
Red lead	2.0		blues
Cobalt oxide	1.5		
	100.0	Color @ C/04	Alice blue (88)
		Color (glaze over)	Medium blue
		Color @ C/8–10	American heritage blue (85)
		Color (glaze over)	Dark blue (93)

S 104 Red Clay

Red clay	85.6	Color @ C/015	Light brick (67)
Crocus martis	8.0	Color (glaze over)	Mottled yellow and brick
Red iron oxide	4.8		
Yellow ochre	1.6		
	100.0	Color @ C/04	Muted orange (68)
		Color (glaze over)	Speckled light brown (37)
		Color @ C/8–10	Metallic gray (116)
		Color (glaze over)	Brown-gray (115)

S 105 Blue-Gray

Tin oxide	85	Color @ C/015	Medium gray (100)
Cobalt oxide	15	Color (glaze over)	Medium gray-blue (91)
	100		
		Color @ C/04	Light blue-gray
		Color (glaze over)	Dark blue (89)
		Color @ C/8–10	Gray-medium blue (91)
		Color (glaze over)	Royal blue (90)

S 106 Gray

Tin oxide	85	Color @ C/015	Gray-pink (45)
Manganese dioxide	7	Color (glaze over)	Mottled dark brown
Red iron oxide	4		and tan
Red copper oxide	3		
Cobalt oxide	1	Color @ C/04	Light gray (104)
	100	Color (glaze over)	Medium gray (116)
		Color @ C/8–10	Medium gray (114)
		Color (glaze over)	Blood red and
			blues

S 107 Blue

Alumina oxide	80	Color @ C/015	Light gray (104)
Zinc oxide	16	Color (glaze over)	Mottled yellow
Cobalt oxide	4		and dark green
	100		
		Color @ C/04	Bright light blue (99)
		Color (glaze over)	Medium blue (93)
		Color @ C/8–10	Bright light blue (99)
		Color (glaze over)	Medium blue (93)

S 108 Greens

Chrome oxide	80	Color @ C/015	Grass green (122)
Flint	13	Color (glaze over)	Mottled red, yellow,
Copper oxide	7		green
	100		
		Color @ C/04	Medium chrome green (122)
		Color (glaze over)	Dark green (129)
		Color @ C/8–10	Medium chrome green (122)
		Color (glaze over)	Mocha red (78)

S 109 Oxide Blues

Zinc oxide	79.1	Color @ C/015	Mouse gray
Cobalt oxide	20.9	Color (glaze over)	Black
	100		
		Color @ C/04	Light green (133)
		Color (glaze over)	Royal blue (90)
		Color @ C/8–10	Medium blue (91)
		Color (glaze over)	Royal blue (90)

S 110 Blue and Green

Alumina oxide	79	Color @ C/015	Mouse gray (100)
Zinc oxide	12	Color (glaze over)	Charcoal gray (115)
Cobalt oxide	9		
	100	Color @ C/04	Bright medium blue (99)
		Color (glaze over)	Dark blue (90)
		Color @ C/8–10	Bright medium blue (90)
		Color (glaze over)	Bright medium blue (90)

S 111 Brown

Zinc oxide	73	Color @ C/015	Muted pink (44)
Red iron oxide	17	Color (glaze over)	Mottled yellow, black
Iron chromate	10		
	100	Color @ C/04	Light tan (39)
		Color (glaze over)	Speckled brown and tan (26)
		Color @ C/8–10	Gray-brown (114)
		Color (glaze over)	Gray to black

S 112 Black

Iron chromate	72	Color @ C/015	Charcoal gray (116)
Cobalt oxide	15	Color (glaze over)	Yellow mixed with
Nickel carbonate (green)	7		black
Tin oxide	6		
	100	Color @ C/04	Charcoal black (115)
		Color (glaze over)	True black (113)
		Color @ C/8–10	Charcoal black (115)
		Color (glaze over)	True black (113)

S 113 Blue or Black

Zinc oxide	71	Color @ C/015	Light blue (88)
Whiting	14	Color (glaze over)	Mottled blue
Cobalt carbonate	11		
Nickel oxide	4	Color @ C/04	Light green (132)
	100	Color (glaze over)	Medium blue (90)
		Color @ C/8–10	*Thin*—charcoal black
			Thick—medium blue
		Color (glaze over)	Medium blue (91)

S 114 Yellows

Tin oxide	71	Color @ C/015	Pale yellow (16)
Whiting	21	Color (glaze over)	Canary yellow (1)
Flint	4		
Chrome potassium	4	Color @ C/04	Pale yellow (16)
	100	Color (glaze over)	Bright light yellow (3)
		Color @ C/8–10	Pale yellow (5)
		Color (glaze over)	Grayed light green (134)

S 115 Buttercup Whites

Tin oxide	70	Color @ C/015	White (32)
Whiting	25	Color (glaze over)	Buttercup (3)
Alumina oxide	5		
	100	Color @ C/04	White (32)
		Color (glaze over)	Off-white (7)
		Color @ C/8–10	White (32)
		Color (glaze over)	Cement white (8)

S 116 Goldenrod

Tin oxide	70	Color @ C/015	Very pale yellow (16)
Whiting	20	Color (glaze over)	Goldenrod (11)
Potassium dichromate	10		
	100	Color @ C/04	Cement white (8)
		Color (glaze over)	Light yellow (3)
		Color @ C/8–10	Light yellow (3)
		Color (glaze over)	Medium moss green (122)

S 117 Neutrals

Albany slip	70	Color @ C/015	Flat brown
Cobalt oxide	8	Color (glaze over)	Black mottled with light green
Copper oxide	8		
Red iron oxide	8		
Manganese dioxide	6	Color @ C/04	Dark gray-brown (146)
	100		
		Color (glaze over)	Blue-black
		Color @ C/8–10	Dark metallic gray (114)
		Color (glaze over)	Blue-black

S 118 Blue-Green

Chrome oxide	69	Color @ C/015	Medium green (139)
Flint	23	Color (glaze over)	Mottled red, yellow,
Cobalt oxide	8		dark green
	100		
		Color @ C/04	Medium muted green (122)
		Color (glaze over)	British green (129)
		Color @ C/8–10	Light turquoise (97)
		Color (glaze over)	British green (130)

S 119 Antimony Oxide

Antimony oxide	67	Color @ C/015	Light yellow (4)
Red lead	25	Color (glaze over)	Buttercup (3)
Zinc oxide	8		
	100	Color @ C/04	White (32)
		Color (glaze over)	Warm white (6)
		Color @ C/8–10	Medium gray (116)
		Color (glaze over)	Medium gray (116)

S 120 Blue-Green

Alumina oxide	66	Color @ C/015	Medium gray (116)
Cobalt oxide	26	Color (glaze over)	Dark green (129)
Chrome oxide	8		
	100	Color @ C/04	Ethan Allen blue (98)
		Color (glaze over)	Blue-black
		Color @ C/8–10	Bright medium blue (95)
		Color (glaze over)	Royal blue (90)

S 121 Gray-Blue

Alumina hydrate	66	Color @ C/015	Light gray (104)
Tin oxide	26	Color (glaze over)	Mottled dark green
Cobalt oxide	5		and yellow
Chrome oxide	2		
Red iron oxide	1	Color @ C/04	Light blue-gray
	100		(102)
		Color (glaze over)	Blue-gray (100)
		Color @ C/8–10	Robin's egg blue (85)
		Color (glaze over)	Dark robin's egg blue (99)

S 122 Browns/Gray

Calcined kaolin	63	Color @ C/015	Mouse gray (100)
Manganese dioxide	19	Color (glaze over)	Dark brown
Iron chromate	16		
Nickel oxide (green)	2	Color @ C/04	Light purplish gray
	100		(100)
		Color (glaze over)	Dark brown (154)
		Color @ C/8–10	Charcoal brown (115)
		Color (glaze over)	Charcoal brown (115)

S 123 Green/Yellow

Tin oxide	60	Color @ C/015	Very pale yellow (16)
Whiting	31	Color (glaze over)	Buttercup (3)
Silica	3		
Lead chromate	6	Color @ C/04	Dirty white (30)
	100	Color (glaze over)	Light violet (47)
		Color @ C/8–10	Light lime green (128)
		Color (glaze over)	Light gray-green (134)

S 124 White

Zircopax	60	Color @ C/015	White (32)
Zinc oxide	20	Color (glaze over)	Light yellow (4)
Whiting	20		
	100	Color @ C/04	White (32)
		Color (glaze over)	Off-white (6)
		Color @ C/8–10	Warm white (15)
		Color (glaze over)	Cement white (8)

S 125 Green

Calcined kaolin	60	Color @ C/015	Muted green (122)
Chrome oxide	20	Color (glaze over)	Mottled red, yellow,
Whiting	20		dark green
	100	Color @ C/04	Muted medium green (122)
		Color (glaze over)	Kelly green (121)
		Color @ C/8–10	Medium gray-green (139)
		Color (glaze over)	Medium gray-green (139)

S 126 White

Alumina oxide	60	Color @ C/015	White (32)
Hydrated borax	20	Color (glaze over)	Light yellow (4)
Magnesium carbonate	10		
Tin oxide	10	Color @ C/04	White (32)
	100	Color (glaze over)	White (7)
		Color @ C/8–10	White (32)
		Color (glaze over)	Cement white (8)

S 127 Turquoise

Copper phosphate	40	Color @ C/015	Light turquoise (133)
Tin oxide	57		
Chrome oxide	3	Color (glaze over)	Mottled dark green and yellows
	100		
		Color @ C/04	Medium gray-brown (146)
		Color (glaze over)	Turquoise (97)
		Color @ C/8–10	Metallic charcoal gray (114)
		Color (glaze over)	Mocha red (78)

S 128 Fawn

Tin oxide	56	Color @ C/015	Muted pink (44)
Lead chromate	16	Color (glaze over)	Mottled yellows, oranges, browns
Iron oxide	14		
Zinc oxide	14		
	100	Color @ C/04	Peach (60)
		Color (glaze over)	Speckled tan (39)
		Color @ C/8–10	Light metallic gray (116)
		Color (glaze over)	Bitter chocolate (153)

S 129 Potpourri

Alumina oxide	55	Color @ C/015	Muted pink (42)
Zinc oxide	20	Color (glaze over)	Buttercup (3)
Red iron oxide	15		
Whiting	10	Color @ C/04	Peach (61)
	100	Color (glaze over)	Light gray (100)
		Color @ C/8–10	Grayed purple (107)
		Color (glaze over)	Grayed green (139)

S 130 Brown

Zinc oxide	53.5	Color @ C/015	Light brownish
Crocus martis	12.5		purple (36)
Chrome oxide	12.5	Color (glaze over)	Mottled yellows,
Lead litharge	9.1		greens, oranges, browns
Borax	9.1		
Red iron oxide	3.3	Color @ C/04	Dusty tan (39)
	100.0	Color (glaze over)	Sweet chocolate (75)
		Color @ C/8–10	Purple-gray (114)
		Color (glaze over)	Bitter chocolate (153)

S 131 Yellow/Gray

Red lead	50	Color @ C/015	Light goldenrod (12)
Antimony oxide	35	Color (glaze over)	Yellow (2)
Tin oxide	15		
	100	Color @ C/04	Clear yellow (3)
		Color (glaze over)	Light yellow-white (5)
		Color @ C/8–10	Steel gray (100)
		Color (glaze over)	Light gray (116)

S 132 Brown

Zinc oxide	50	Color @ C/015	Medium brown (36)
Potassium chrome	35	Color (glaze over)	Mottled yellows,
Red iron oxide	15		oranges, browns
	100	Color @ C/04	Light brown
		Color (glaze over)	Speckled brown and yellow (22)
		Color @ C/8–10	Charcoal gray (114)
		Color (glaze over)	Greenish brown (155)

S 133 Blue/Gray

China clay	50	Color @ C/015	Medium purple-brown (36)
Black iron oxide	28	Color (glaze over)	Mottled black, greens
Cobalt oxide	14		
Manganese dioxide	8		
	100		
		Color @ C/04	Deep purple-gray (107)
		Color (glaze over)	Royal blue (90)
		Color @ C/8–10	Mouse gray (116)
		Color (glaze over)	Regent's blue (89)

S 134 Green

Zinc oxide	50	Color @ C/015	Gray-green (134)
Flint	25	Color (glaze over)	Mottled greens, yellows
Chrome oxide	25		
	100		
		Color @ C/04	Gray-green
		Color (glaze over)	Olive green (138)
		Color @ C/8–10	Medium green (122)
		Color (glaze over)	Moss green (141)

S 135 Yellow

Red lead	50	Color @ C/015	Light goldenrod (12)
Antimony oxide	25	Color (glaze over)	Yellow (2)
Zinc oxide	25		
	100	Color @ C/04	Medium yellow (12)
		Color (glaze over)	Warm white (15)
		Color @ C/8–10	Confederate gray (100)
		Color (glaze over)	Confederate gray (100)

S 136 Green

Tin oxide	50	Color @ C/015	Light lime green
Whiting	25		(127)
Silica	10	Color (glaze over)	Mottled oranges,
Borax	10		yellows, greens
Chrome oxide	5		
	100	Color @ C/04	Lime sherbet (127)
		Color (glaze over)	Glen green (143)
		Color @ C/8–10	Gray-green (143)
		Color (glaze over)	Pale green (135)

S 137 Green

China clay	50	Color @ C/015	Light gray-green
Alumina hydrate	22		(134)
Chrome oxide	12	Color (glaze over)	Dark green
Flint	10		
Cobalt oxide	6	Color @ C/04	Medium gray-green
	100		(134)
		Color (glaze over)	Dark turquoise (129)
		Color @ C/8–10	Warm blue (83)
		Color (glaze over)	Light gray (104)

S 138 Brown

Zinc oxide	50	Color @ C/015	Medium brown (36)
Red iron oxide	20	Color (glaze over)	Mottled dark brown,
Whiting	10		yellow, green, orange
Alumina hydrate	10		
Chrome oxide	10	Color @ C/04	Toast brown (37)
	100	Color (glaze over)	Rich brown (75)
		Color @ C/8–10	Charcoal gray (114)
		Color (glaze over)	Toast brown (37)

S 139 Potpourri

Zinc oxide	50	Color @ C/015	Muted pink (44)
Tin oxide	20	Color (glaze over)	Mottled yellows
Whiting	10		
Red iron oxide	10	Color @ C/04	Light tan (40)
China clay	10	Color (glaze over)	Light gray (104)
	100		
		Color @ C/8–10	Dark metallic brown (73)
		Color (glaze over)	Light gray

S 140 Green

Tin oxide	48.5	Color @ C/015	Light lime green (128)
Whiting	24.8		
Flint	17.5	Color (glaze over)	Yellow (2)
Borax	3.9		
White lead	3.9	Color @ C/04	Very light tan (29)
Chrome oxide	1.4	Color (glaze over)	Light gray-green
	100.0		
		Color @ C/8–10	Medium gray-green (139)
		Color (glaze over)	Pale green (135)

S 141 Yellow

Zircopax	48	Color @ C/015	Light yellow (14)
Flint	31	Color (glaze over)	Buttercup (3)
Vanadium pentoxide	13		
Barium carbonate	8	Color @ C/04	Light yellow (4)
	100	Color (glaze over)	Off-white (31)
		Color @ C/8–10	White (7)
		Color (glaze over)	Warm tan

S 142 Goldenrod

Red lead	47	Color @ C/015	Goldenrod (12)
Antimony oxide	28	Color (glaze over)	Gold (13)
Feldspar	13		
Tin oxide	6	Color @ C/04	Goldenrod (12)
Lead chromate	6	Color (glaze over)	Light gray
	100		
		Color @ C/8–10	Light tan (28) and medium gray (100)
		Color (glaze over)	Cement white (8)

S 143 Gray

Nepheline syenite	40	Color @ C/015	Light gray (151)
Flint	38	Color (glaze over)	Medium olive green (138)
Nickel oxide	20		
Cobalt oxide	2		
	100	Color @ C/04	Light gray (104)
		Color (glaze over)	Medium gray
		Color @ C/8–10	Light charcoal-purplish-gray (114)
		Color (glaze over)	Light charcoal-purplish-gray (114)

S 144 Brown

Zinc oxide	40	Color @ C/015	Muted purple (109)
Whiting	20	Color (glaze over)	Dark brown (154)
Calcined kaolin	20		
Iron oxide	10	Color @ C/04	Toast brown (39)
Manganese dioxide	10	Color (glaze over)	Medium brown to black
	100		
		Color @ C/8–10	Medium gray (116)
		Color (glaze over)	Light tan (28)

S 145 Pink

Tin oxide	36.4	Color @ C/015	Hint of pink (63)
Whiting	22.7	Color (glaze over)	Light yellow (4)
Flint	20.0		
Cerium oxide	11.8	Color @ C/04	Peach ice cream (46)
Fluorspar	9.1	Color (glaze over)	Hint of pink (63)
	100.0		
		Color @ C/8–10	Not determined

S 146 Gray/Blue

Zinc oxide	35.6	Color @ C/015	Mouse gray (100)
Cornwall stone	35.6	Color (glaze over)	Dark brown
Manganese oxide	17.7		
Cobalt oxide	11.1	Color @ C/04	Medium gray-blue (116)
	100.0		
		Color (glaze over)	Dark blue (90)
		Color @ C/8–10	Charcoal gray (114)
		Color (glaze over)	Royal purple (119)

S 147 Gray II

Black soda uranium	35	Color @ C/015	Light gray-green (136)
E.P. kaolin	35		
Silica	20	Color (glaze over)	Medium goldenrod (11)
Tin oxide	10		
	100		
		Color @ C/04	Cement white (8)
		Color (glaze over)	Pale yellow (14)
		Color @ C/8–10	Speckled gray (151)
		Color (glaze over)	Light gray (104)

S 148 Gray/Green

Rutile	35	Color @ C/015	Medium gray (151)
Zinc oxide	30	Color (glaze over)	Medium green (142)
Tin oxide	20		
Copper carbonate	15	Color @ C/04	Light tan (28)
	100	Color (glaze over)	Turquoise (139)
		Color @ C/8–10	Charcoal gray (114)
		Color (glaze over)	Rust and gray-green (76 and 143)

S 149 Gray III

Copper carbonate	33.4	Color @ C/015	Deep gray-purple (107)
Iron oxide	33.3		
Zircopax	33.3	Color (glaze over)	Mottled yellow/ dark turquoise
	100.0		
		Color @ C/04	Charcoal gray (114)
		Color (glaze over)	Dark green (129)
		Color @ C/8–10	Medium charcoal gray (116)
		Color (glaze over)	Reddish brown (49)

S 150 Potpourri II

Zinc oxide	32	Color @ C/015	Dirty khaki
China clay	28	Color (glaze over)	Mottled yellow, green, orange
Chrome oxide	22		
Iron oxide	18		
	100	Color @ C/04	Tannish green (159)
		Color (glaze over)	Medium brown (156)
		Color @ C/8–10	Dark grayish purple (107)
		Color (glaze over)	Medium gray (146)

S 151 Green/Yellow

Potassium dichromate	30	Color @ C/015	Very pale lime (136)
Flint	30	Color (glaze over)	Medium goldenrod
Whiting	20		(11)
Fluorspar	20		
	100	Color @ C/04	Pale green (127)
		Color (glaze over)	Buttercup yellow (2)
		Color @ C/8–10	Cool brownish green (155)
		Color (glaze over)	Fawn green (143)

S 152 Butterscotch

Rutile	30	Color @ C/015	Muted pink (44)
Flint	30	Color (glaze over)	Mottled yellows
Zinc oxide	20		
Tin oxide	10	Color @ C/04	Butterscotch (62)
Red iron oxide	10	Color (glaze over)	Butterscotch (62)
	100		
		Color @ C/8–10	Indian brown (49)
		Color (glaze over)	Dark gray-green (155)

IV

Earthenware
Leadless Glazes

Glazes come in many types, colors, firing temperatures, and surfaces; yet they all may be defined as a layer of glass fused to a clay form. *Glass* is a supercooled liquid of high viscosity at ordinary room temperatures. A glaze is a special glass comprising inorganic oxides, of which silica and fluxes are the most important. The mineral oxides are applied in powder form to a ceramic body by spraying, painting, dipping, or pouring. It is during the firing that the individual minerals coalesce and fuse to the clay body. The compact molten surface, upon cooling, does not separate into its original minerals. The glaze covering imparts the following qualities to the clay surface:
1. Resistance to scratching and abrasion
2. Imperviousness to household acids and alkalies
3. Insolubility in water and other liquids (making the clay container a water-tight vessel)
4. Decorative and aesthetic qualities
5. Surface quality (glossy, semiglossy, matt, satin, etc.)
6. Color
7. Hard, smooth, durable surface that is easy to clean

Problems with Glazes

Many factors, including glaze thickness, composition of the clay body, impurities in the clay body and glaze, length of soaking time, maturing temperature, types of glazes on surrounding clay bodies in the kiln, character of the firing atmosphere, type of fuel burned, and colorants used, determine how the glaze—both leadless and leaded—will respond. Should the glaze be applied too thin, for example, the result will be a rough surface because of the insufficient amount of glaze materials available to complete the glaze coalescence. Moreover, impurities in the clay, such as iron, will impart color or will affect the desired glaze quality. If soaking at peak temperature is permitted, it will cause bubbles to disperse, the glaze to mature, and pits and glaze application marks to heal. Underfiring or overfiring the glaze may result in undesirable surface qualities. Although some glazes have a several-cone firing range, others have a precise fusing temperature that must be closely watched. Finally, a reductive or oxidative kiln atmosphere can blemish or alter the glaze, as can a neutral atmosphere that contains the impurities of dust, ash, sulphur, or fumes migrating from adjacent pots.

To minimize the negative influence of some of these factors, the ceramist must carefully monitor the mixing, application, and firing procedure of a glaze. Keeping accurate records of the procedures will be invaluable for correcting faulty glazes, improving adequate ones, and, in general, for maintaining consistent and dependable glazes. Should a glaze not have all the desired properties, then the formula, the application method, and/or the firing procedure needs to be altered. The potter must be cautious in modifying formulas, though. While it is possible to arrive at the same molecular formula by substituting one mineral for another, the results produced may not be the same. Exchanging the feldspar for another mineral, for example, may slightly modify the appearance of the surface, which may become a little more fluid or a little less transparent. Often, just adding or subtracting a small amount of one mineral in the formula can alter it enough to improve the glaze. In any event, whenever a new or an altered formula is to be tried, it should first be tested.

Recommended Glaze-Testing Procedure

Testing a glaze is a quick process and well worth the time spent. First, 100 grams of the glaze is mixed in a 6-ounce paper cup. Second, the bisque-fired test bar is dipped into the glaze until the vertical and horizontal surfaces are covered, and then a second dip is made half-way down the vertical surface (Drawing 6). Third, the test piece is fired to the recommended temperature. This simple procedure will provide a great deal of information about the glaze.

The author has tested several thousand ceramic formulas, of which only 40 percent or fewer have proved to be desirable. Many of the formulas have no particular personality, or simply do not work. The clay test piece for glazes used by the author has evolved during testing over the years. The first test piece was a clay bar 2 inches long, then came a 4-inch bar, a 4-inch slightly bent bar, wheel-thrown pie-shaped bars, a 4-inch bar bent to an L shape, a T shape bar, and, finally, the extruded shape of the simple L that has become the more sophisticated one shown in Figure 10. This L shape bar, 2½ inches high, is cut in an aluminum templet. Several yards of extruded clay are placed on wareboards, and uniform test pieces are cut. Several hundred bars can be made in minutes. When dry, the brown clay bars are dipped half-way into a white engobe to provide contrast for glaze testing. The cupped edges of the test piece provide a catch basin for fluid glazes. The L shape test piece will provide many bits of information in one simple glaze test, including the following:

Horizontal and vertical surface glaze quality
Color
Texture
Degree of glaze flow
Ability of undercolors to show through
Transparency
Glaze fit
Difference between one-glaze and two-glaze thickness
Pooling effect in the scalloped (finger) ridges
Quantity of glaze adhering to top edge

Drawing 6
Glaze Test Piece

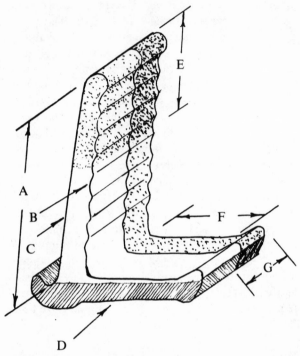

A. Test bar painted or dipped into glaze for one coat
B. Scalloped edge to simulate finger grooves of a thrown pot
C. Flat surface for vertical test
D. Identification key put on bottom
E. Test bar dipped or painted for a second coat
F. Flat surface for horizontal test
G. Test bar dipped halfway into contrasting-color engobe before bisque-firing

Leadless Glazes

Lead glazes, which have been used since ancient times, have the desirable features of ease of application, even flow, bright colors, and a very smooth, blemish-free surface. Because of the potential toxicity of lead, however, potters and the ceramic industry have long been interested in finding low-temperature leadless glazes. In northern Europe during the late nineteenth century, the concern over lead poisoning stimulated the development of modified-lead, leadless, high-zinc (bristol), and other glazes. No serious alternative to lead was found, though, until the technique of fritting was perfected. If done properly, fritting renders the lead nontoxic. Nevertheless, the public still reacts negatively to the idea of a glaze containing lead, even if the amount is low and the glaze has been certified safe. (In the United States, some government regulation of lead content has been enacted; in Britain, a glaze is classified as leadless if the lead content is less than 1 percent.)

EARTHENWARE LEADLESS-GLAZE FORMULAS

Many attractive colors and finishes are possible in leadless glazes. Lithium blue, coral red, and uranium yellow are examples. In this chapter, a variety of glaze formulas are listed, some with raw mixtures, some with leadless frits, and others with frits bearing a small amount of lead. The maturing temperatures range from C/015 through C/04, with surfaces of gloss, semigloss, and matt. At low temperatures, the fluxing agents used are soda, boron, zinc, and lithium. Some formulas contain colorants.

CONE/015

G 3001 Slight White

Frit #3110 FERRO				
(leadless)			Temperature	C/015
		100	Surface @ C/015	Gloss
		100	Fluidity	None
			Stain penetration	All
			Opacity	Translucent
			Color / oxidation	Slight white (16)
			NOTE: Some cracks	

G 3002 Cream-Yellow

Frit #3110 FERRO		Temperature	C/015
(leadless)	92.0	Surface @ C/015	Semigloss
Tennessee ball	5.5	Fluidity	None
Lead chromate	1.5	Stain penetration	All
Tin oxide	1.0	Opacity	Translucent
	100.0	Color / oxidation	Cream-yellow (4)

G 3003 Transparent

Frit #25 (leadless)	65	Temperature	C/015
Lithium carbonate	16	Surface @ C/015	Gloss
Gerstley borate	16	Fluidity	None
Tennessee ball #1	3	Stain penetration	All
	100	Opacity	Transparent
		Color / oxidation	Clear
		NOTE: Some cracks	
Yellow ochre	7	Color / oxidation	Light brown (29)
Uranium	6	Color / oxidation	Slight yellow (4)
Chrome oxide	3	Color / oxidation	Bright chrome green (129)
Nickel carbonate	3	Color / oxidation	Light gray-brown (36)

G 3004 Semigloss

Frit #25 PEMCO	55	Temperature	C/015
Colemanite	45	Surface @ C/015	Semigloss
	100	Fluidity	Little
		Stain penetration	All
		Opacity	Transparent when thin Translucent when thick
		Color / oxidation	Slight white (16)

G 3005 Leadless

Frit #25 PEMCO	44	Temperature	C/015
Frit #3124 FERRO		Surface @ C/015	Semigloss
(leadless)	26	Fluidity	None
Flint	14	Stain penetration	All
Lithium carbonate	10	Opacity	Transparent
Georgia china clay	6	Color / oxidation	Clear
	100	NOTE: Some cracks	
Orange ferrite	3	Color / oxidation	Tan (38)

G 3006 Gloss

Frit #3230 FERRO		Temperature	C/015
(leadless)	43.5	Surface @ C/015	Gloss
Frit #25 PEMCO	53.1	Fluidity	None
Lithium carbonate	3.4	Stain penetration	All
	100.0	Opacity	Transparent
Bentonite	1.0	Color / oxidation	Slight white (16)
Antimony	8.0 ⎫	Opacity	Opaque
Cobalt carbonate	0.5 ⎭	Color / oxidation	Medium blue (59)

G 3007 Water Clear

Gerstley borate	42.4	Temperature	C/015
Frit #25 PEMCO	25.3	Surface @ C/015	Gloss
Flint	18.2	Fluidity	Little
Cryolite	9.1	Stain penetration	All
Lithium carbonate	5.0	Opacity	Transparent
	100.0	Color / oxidation	Water clear
		NOTE: Some cracks	
Chrome oxide	1 ⎫		
Yellow base (ochre)	6 ⎬ Opacity		Opaque
Tin oxide	2 ⎭ Color / oxidation		Grass green (125)

CONE/014

G 3008 High Gloss

Frit #5301 (leadless)	55	Temperature	C/014
Barium carbonate	15	Surface @ C/014	High gloss
Lithium carbonate	15	Fluidity	Some
Whiting	6	Stain penetration	All
Georgia china clay	6	Opacity	Transparent
Flint	3	Color / oxidation	Water clear
	100	NOTE: Some cracks	

CONE/012

G 3009 Transparent Matt (Lead Frit)

Frit #25 PEMCO	50	Temperature	C/012
Frit #3304 FERRO	35	Surface @ C/012	Matt
Whiting	15	Fluidity	None
	100	Stain penetration	All
		Opacity	Transparent
		Color / oxidation	Clear
Lead chromate (orange)	3	Color / oxidation	Bright lemon yellow (2)
Chrome oxide	2 ⎫	Color / oxidation	Dull green–green
Tin oxide	4 ⎭		(139)

CONE/010

G 3010 Transparent High Gloss

Frit #5301 FERRO	100	Temperature	C/010
	100	Surface @ C/010	High gloss
		Fluidity	Little
		Stain penetration	All
		Opacity	Transparent
		Color / oxidation	Clear
		NOTE: Some cracks	
Silver nitrate	3 ⎫	Color / oxidation	Clear, slight silver luster
Soda ash	1 ⎭		
Tin oxide	10	Color / oxidation	White (32)

G 3011 Leadless

Frit #14 HOMMEL (leadless)	81	Temperature	C/010
Georgia kaolin	10	Surface @ C/010	Semi-matt
Whiting	9	Fluidity	Little
	100	Stain penetration	All
		Opacity	Transparent
		Color / oxidation	Clear
Zinc oxide	6	Opacity	Translucent
Tin oxide	16	Color / oxidation	Light green (135)
Antimony oxide	22		

G 3012 Leadless Bright Gloss

Frit #24 PEMCO	56	Temperature	C/010
Lithium carbonate	19	Surface @ C/010	Bright gloss
Uranium (black)	17	Fluidity	Fluid
Magnesium carbonate	5	Stain penetration	None
Tin oxide	2	Opacity	Translucent when thin
Bentonite	1		Opaque when thick
	100	Color / oxidation	Bright rich brown (154)

CONE/06

G 3013 Clear-Clear

Frit #25 PEMCO	100	Temperature	C/06
	100	Surface @ C/06	Brilliant gloss
Bentonite	2	Fluidity	None
		Stain penetration	All
		Opacity	Transparent
		Color / oxidation	Clear-clear

G 3014 Leadless Gloss

Frit #14 HOMMEL	90	Temperature	C/06
Tennessee ball #1	10	Surface @ C/06	Gloss
	100	Fluidity	Little
		Stain penetration	All
		Opacity	Transparent
		Color / oxidation	Clear
		NOTE: Some cracks	

G 3015 Leadless Frit

Frit #54 PEMCO	85	Temperature	C/06
Spodumene	15	Surface @ C/06	Gloss
	100	Fluidity	Little
		Stain penetration	All
		Opacity	Transparent
		Color / oxidation	Clear
		NOTE: Some cracks	

G 3016 Milky Transparent

Gerstley borate	70	Temperature	C/06
Flint	25	Surface @ C/06	Semigloss
Georgia kaolin	5	Fluidity	None
	100	Stain penetration	All
		Opacity	Transparent
		Color / oxidation	Milky

G 3017 Clear Satin Matt

Frit #14 HOMMEL	44	Temperature	C/06
Tennessee ball #1	28	Surface @ C/06	Satin matt
Wollastonite	28	Fluidity	None
	100	Stain penetration	All
		Opacity	Transparent
		Color / oxidation	Clear
Burnt sienna	4	Color / oxidation	Light tan
Cobalt	0.6 ⎫	Color / oxidation	Medium steel blue
Red iron	1.8 ⎭		(92)
Nickel carbonate	2.5 ⎫	Color / oxidation	Gray-yellow-
Copper carbonate	0.5 ⎭		green (152/149)

G 3018 Alfred University

Flint	39.7	Temperature	C/06
Potassium nitrate	21.9	Surface @ C/06	Gloss/matt
Soda ash	11.5	Fluidity	Fluid
Whiting	10.5	Stain penetration	None
Black copper oxide	8.9	Opacity	Opaque
Boric acid	4.7	Color / oxidation	Purple matt over
Magnesium carbonate	2.8		glossy blues (118/93)
	100.0		

CONE/04

G 3019 Glossy Tans

Lead monosilicate	69	Temperature	C/04
Nepheline syenite	20	Surface @ C/04	Gloss
Whiting	9	Fluidity	Fluid
Bentonite	2	Stain penetration	All
	100	Opacity	Transparent
		Color / oxidation	Light tan (38)
Chrome oxide	1.5	Color / oxidation	Broken light greens and tans (126/38)

G 3020 Van Winkle: Alkaline Copper

Frit #3110 FERRO	55	Temperature	C/04
Flint	20	Surface @ C/04	Gloss
Frit #5301 FERRO	15	Fluidity	Little
Tennessee ball #1	10	Stain penetration	All
	100	Opacity	Transparent
		Color / oxidation	Clear
		NOTE: Some cracks	

G 3021 Dennis's Lithium Blue

Flint	53.9	Temperature	C/04
Lithium carbonate	27.0	Surface @ C/04	Matt
Georgia kaolin	13.6	Fluidity	None
Bentonite	2.8	Stain penetration	All
Copper carbonate	2.6	Opacity	Translucent
	99.9	Color / oxidation	Blue/Green (93/97)

G 3022 Newcomb Matt Leadless

Frit #2106 HOMMEL	47.0	Temperature	C/04
Oxford feldspar	23.2	Surface @ C/04	Gloss/semigloss
Georgia kaolin	16.5	Fluidity	None
Whiting	5.5	Stain penetration	Darks
Flint	5.0	Opacity	Opaque
Tin oxide	2.8	Color / oxidation	White (7)
	100.0		

G 3023 Leadless Satin Matt

Flint	42	Temperature	C/04
Georgia kaolin	22	Surface @ C/04	Satin matt
Lithium carbonate	10	Fluidity	None
Zinc oxide	10	Stain penetration	None
Whiting	10	Opacity	Opaque
Barium carbonate	6	Color / oxidation	Charcoal (114)
	100		
Manganese carbonate	3	Color / oxidation	Brown (153)
Copper carbonate	3	Color / oxidation	Deep yellow-green (126)

G 3024 Leadless Translucent

Fritt #3124 FERRO	40	Temperature	C/04
Cullet glass	40	Surface @ C/04	Matt
Tennessee ball #1	10	Fluidity	None
Flint	10	Stain penetration	All
	100	Opacity	Translucent
		Color / oxidation	Milky / clear
Crocus martis	6 ⎱	Color / oxidation	Dark purple (117)
Ilmenite	1 ⎰		with specks
Kryolith	9 ⎱	Color / oxidation	Dark purple and
Magnetite	5 ⎰		tan-gray (117/151)
Raw sienna	3	Color / oxidation	Tan-gray (151)
Rutile	4 ⎱	Color / oxidation	Tan (29)
Burnt umber	4 ⎰		

G 3025 Leadless Medium-Dark Blue

Calcined borax	37.0	Temperature	C/04
Silica	25.0	Surface @ C/04	Gloss
Oxford feldspar	21.0	Fluidity	None
Whiting	11.0	Stain penetration	Darks
Silica	4.0	Opacity	Translucent
Cobalt carbonate	1.2	Color / oxidation	Medium-dark blue
Red iron oxide	0.8		(91)
	100.0		

V

Singlefire Glazes

Singlefiring, also known as *slipglazing, once-firing,* and *green-glazing,* is the technique of applying a glaze to raw (unfired) clay ware and firing it only once, to the maturing temperature. Wares from all over the world, including burnishedware, wood-ash-glazed ware, tenmokuware, salt-glazed ware, the Roman terra sigillata pottery, and slipglazed ware—as well as many terra-cotta, stoneware, and earthenware sculpture—are made by the singlefire method. The technique of singlefiring, moreover, was the first firing method employed by potters, and it is still widely used today. Because singlefiring requires only one loading and firing of the kiln—the bisque firing is omitted—both time and money (particularly in fuel costs) can be saved.

Although there are many advantages to singlefiring, there can also be problems. Depending on the glaze used, difficulties that might arise include bubbling of the glaze; flaking away of the glaze from the pot because of the shrinkage differential; saturation of the clay with water during the glaze application, which causes thin-walled pots to collapse; and deformation of fine detail of the greenware as the glaze is painted. Also, some types of singlefire glazes have limited color ranges.

The glazes most frequently used in singlefiring are ash, terra sigillata, slip, and glazes with a high clay content. In general, the glaze or glazes can be applied to wet, leather-hard, and dry greenware, but in selecting a glaze the ceramist must take into account the extent of clay shrinkage that will occur in drying and firing. Most slip glazes will adhere to both wet and dry greenware, but ash and high-clay-content glazes are applied only to dry greenware. Some regular glazes will also work well as singlefire glazes; testing will determine if they will flake off during drying and firing (see page 125).

A singlefire glaze can be applied by spraying, dipping, painting, splattering, or pouring. For subtle texture and design on the clay form, spraying is recommended. The glaze is used on the inside of pots for decoration, color, and smooth finish. The insides are glazed first and set to dry; then the outsides are glazed. Singlefire glazeware is even more delicate than greenware. The ware must be brought up through the drying stages in the firing very carefully. If steam and carbonaceous matter are permitted to escape too quickly, the glaze will flake off. The firing in this part of the cycle is slower than typical bisque firing.

Ash Glazes

Wood ash was one of the first materials developed for high-fire ceramic glaze. Its use as a glaze was discovered accidentally when, during firing, ash from the wood fire drifted into the kiln chamber and landed on a pot. The fluxes, magnesia, and calcium in the ash fused with the silica in the clay to form a glaze on the shoulders and exposed areas of the pot. Then, early in the historical development of high-fire glazes came the discovery that ash mixed with a small amount of clay could be easily applied to a piece of pottery to produce a glaze.

Ash glazes are generally derived from wood, but straw, grass, or other organic material, when burned, will leave a powdery residue that can also serve as an ash glaze base. The mineral content of different ashes will vary enormously. In an apple tree, for example, the trunk wood has more calcium than the bark; the bark of new or young growth has less silica than old bark; the wood in spring has more calcium and alkalies; and apple trees growing on the side of a hill have lower mineral content than those in the valley. Some of the most interesting ash glazes that can be used (and that can be obtained easily and inexpensively) are fireplace and incinerator ash. Because of the variety of items which are burned and become part of the ash, one never knows what sort of glaze may result. Whatever kind of ash is used, however, it must be screened several times to remove carbon and unburned materials.

Ash contains soda, potash, magnesia, calcium, silica, phosphorus pentoxide, alumina, and traces of metallic oxides. The phosphorus pentoxide and the metallic oxides give the glaze its opacity, flecking, and subtle colors. Some ashes are a complete glaze and are used as such, especially at C/10 or higher.

A basic wood ash glaze for C/8–9 is as follows:

Washed and screened ash, 60 mesh	60
Flux (colemanite, spar, etc.)	30
Ball clay (binder)	10
	100
Copper or other colorant	2

Since most ashes will not make a complete glaze at temperatures under C/10, however, the addition of a flux—like colemanite, feldspar, nepheline syenite, or gerstley borate—together with a clay or bentonite binder, is usually necessary. Below C/02, the use of ash in a glaze is pointless.

Terra Sigillata

Terra sigillata is a type of glaze found on ancient pottery having a red-brown color, a fine-textured body, and a slightly glossy sheen. The best known of the ancient Roman ceramics that bear raised decoration were created with terra sigillata. The Romans used a clay deposit from Lemnos and Samos, but local red earthenware can be used. The earthenware clay is mixed for several hours with a large quantity of water; a ball mill can be used. Small amounts of potash, soda, zinc, and/or rutile are added to the mixing clay. The mix is permitted to settle, the water is decanted, and the top 1 inch of the mixture is used. It is the fine particles of clay and traces of flux that make up the slip, which is applied as a thin layer to greenware. Earthenware temperatures and mid-range temperatures are the most common ones used.

Slip-Glazing

Slip glazing, a natural clay containing flux and iron, can be applied to greenware and leather-hard and dry clay forms. Albany slip is the best known and, used by itself, creates a dark brown, glossy, translucent (when applied thickly) C/10 glaze. Barnard clay, as well as red and brown earthenware clay bodies when mixed with flux, will also produce a glaze. The amount of flux to be added depends upon the clay type and the firing temperature. Slip glazes, Albany in particular, will fire to a smooth finish, with few pinholes, and are hard and durable. In general, because of the iron in the clays, the colors produced range from medium tan to dark brown. By reducing the amount of Albany or other clays in the formula, the ceramist can create some other colors and surfaces, but the possibilities are still limited.

Cost

With the ever-increasing price of fuel, the use of singlefire glazes is becoming more popular. Cost saving is one of the major reasons why many potters are now using salt and/or soda glazes in firing. The cost of 100 pounds of slip glaze is about 30 cents per pound, while 100 pounds of a low-cost stoneware is about 15 cents per pound. However, a high percentage of zinc, zircopax, cobalt, spodumene, rutile, or cornwall stone minerals added to a stoneware glaze could very easily push the cost up to 39 cents per pound. The real economy is not so much in the glaze cost, though, but rather in the savings of time and fuel realized from the singlefire method.

Testing Singlefire Glazes

Many formulas were selected for testing. Each was weighed to 100-gram units and mixed with a sufficient amount of water to produce a dipping consistency. Then two test pieces were dipped into each mixture. One test piece was a bisque-fired brown stoneware dipped half-way into a white engobe, and the other a dry greenware porcelain clay (see Drawing 6, page 112). Both test pieces were fired to the appropriate temperature and the results of each test were recorded. The author, besides seeking typical information on glazes, was trying to find out how well the slip glaze adhered to the dry greenware and whether the glaze peeled, scaled, or flaked off the test piece. If any of these events did occur, the glazes responsible were omitted. The porcelain piece was fired in a gas kiln to maturing temperature and reduced. The stoneware piece was fired in an electric kiln to its maturing temperature. Formulas were tested at various temperatures, producing different results.

SINGLEFIRE-GLAZE FORMULAS

CONE/3

SG 301 Albany Slip

Albany slip	90	Temperature	C/3
Gerstley borate	10	Surface @ C/3	Smooth matt
	100	Fluidity	Little
		Stain penetration	Darks
		Opacity	Opaque
		Color / oxidation	Olive mustard (147)

SG 302 Broken Matts

Albany slip	60	Temperature	C/3–5
Whiting	16	Surface @ C/3	Broken matts
Zinc oxide	10	Fluidity	Fluid
Spodumene	7	Stain penetration	None
Red iron oxide	5	Opacity	Opaque
Cobalt oxide	2	Color / oxidation	Broken black and
	100		olive (129/142)

SG 303 Brown Matt

Albany slip	36.4	Temperature	C/3
Ball clay	22.7	Surface @ C/3	Matt
Wallastonite	18.2	Fluidity	None
Nepheline syenite	13.6	Stain penetration	None
Red iron oxide	9.1	Opacity	Opaque
	100.0	Color / oxidation	Charcoal brown (155)

CONE/4

SG 304 Transparent Olive

Albany slip	60	Temperature	C/4
Colemanite	20	Surface @ C/4	Semigloss
Cornwall stone	20	Fluidity	Little
	100	Stain penetration	All
		Opacity	Transparent
		Color / oxidation	Slight olive (150)

SG 305 High-Gloss Oaktag

Tennessee ball	45	Temperature	C/4
Sodium bicarbonate	20	Surface @ C/4	High gloss
Gerstley borate	20	Fluidity	None
Borax	10	Stain penetration	All
Zinc oxide	5	Opacity	Transparent
	100	Color / oxidation	Oaktag (15)

SG 306 High-Gloss Medium Brown

Barnard clay	42	Temperature	C/4–6
Sodium bicarbonate	33	Surface @ C/4	High gloss
Zircopax	9	Fluidity	Very fluid
Lithium	8	Stain penetration	None
Talc	8	Opacity	Opaque
	100	Color / oxidation	Medium brown (153)

CONE/5

SG 307 Opaque Army Green

Albany slip	90	Temperature	C/5
Zircopax	10	Surface @ C/5	High gloss to
	100		satin matt
		Fluidity	Little
		Stain penetration	Darks
		Opacity	Opaque
		Color / oxidation	Army green (137)

SG 308 Medium Gray

Albany slip	90	Temperature	C/5
Titanium dioxide	10	Surface @ C/5	Matt
	100	Fluidity	Some
		Stain penetration	None
		Opacity	Opaque
		Color / oxidation	Medium gray (116)

SG 309 Slip Black

Barnard clay	90	Temperature	C/5
Whiting	10	Surface @ C/5	Matt
	100	Fluidity	Little
		Stain penetration	None
		Opacity	Opaque
		Color / oxidation	Black (115)

SG 310 Slip Brown

Albany slip	70	Temperature	C/5
Magnesium carbonate	30	Surface @ C/5	Broken gloss/semi-
	100		gloss
		Fluidity	Fluid
		Stain penetration	Darks
		Opacity	Opaque when thick
			Translucent when thin
		Color / oxidation	Brown (73)

SG 311 Olive Green

Albany slip	70	Temperature	C/5
Cryolite	30	Surface @ C/5	Gloss/high gloss
	100	Fluidity	Fluid
		Stain penetration	Most
		Opacity	Translucent
		Color / oxidation	Olive green (147)

SG 312 Smooth Matt

Albany slip	60	Temperature	C/5
Rutile	40	Surface @ C/5	Smooth matt
	100	Fluidity	No
		Stain penetration	Darks
		Opacity	Opaque
		Color / oxidation	Medium tan (35)

SG 313 Gray-Green

Albany slip	60	Temperature	C/5
Wollastonite	40	Surface @ C/5	Smooth satin matt
	100	Fluidity	Some
		Stain penetration	Most
		Opacity	Translucent
		Color / oxidation	Gray-green (138)

SG 314 Gray-Brown

Albany slip	60	Temperature	C/5
Nepheline syenite	40	Surface @ C/5	Semigloss
	100	Fluidity	None
		Stain penetration	Darks
		Opacity	Opaque
		Color / oxidation	Gray-brown (155)

SG 315 Dark Charcoal Gray

Albany slip	60	Temperature	C/5
Burnt umber	40	Surface @ C/5	Semigloss
	100	Fluidity	Little
		Stain penetration	None
		Opacity	Opaque
		Color / oxidation	Dark charcoal gray (114)

Note: Separates when thick

SG 316 Matt Black

Albany slip	60	Temperature	C/5
Nepheline syenite	30	Surface @ C/5	Matt
Yellow ochre	10	Fluidity	Little
	100	Stain penetration	None
		Opacity	Opaque
		Color / oxidation	Black (113)

SG 317 Broken Dark Browns

Barnard clay	60	Temperature	C/5
Whiting	20	Surface @ C/5	Broken gloss/matt
Gerstley borate	20	Fluidity	Some
	100	Stain penetration	None
		Opacity	Opaque
		Color / oxidation	Broken dark browns (74)

SG 318 Purple/Black

Albany slip	60	Temperature	C/5
Cornwall stone	20	Surface @ C/5	Matt
Red iron oxide	20	Fluidity	Little
	100	Stain penetration	None
		Opacity	Opaque
		Color / oxidation	Purple/black (119/115)

SG 319 Semimatt

Albany slip	60	Temperature	C/5
Cornwall stone	20	Surface @ C/5	Semimatt
Whiting	10	Fluidity	Some
Red iron oxide	10	Stain penetration	None
	100	Opacity	Opaque
		Color / oxidation	Blacks (113) when thick
			Olives (138) when medium
			Rust (50) when thin

SG 320 Volcanic Matt

Barnard clay	50	Temperature	C/5–6
Flint	24	Surface @ C/5	Volcanic matt
Wollastonite	10	Fluidity	None
Frit #14 HOMMEL	10	Stain penetration	None
Fluorspar	6	Opacity	Opaque
	100	Color / oxidation	Dark browns (74)

CONE/6

SG 321 Chocolate Brown Slip

Albany slip	90	Temperature	C/6–7
Yellow ochre	10	Surface @ C/6	High gloss
	100	Fluidity	Little
		Stain penetration	None
		Opacity	Opaque
		Color / oxidation	Chocolate brown (74)

SG 322 Transparent Browns

Albany slip	90	Temperature	C/6
Barium carbonate	10	Surface @ C/6	Gloss
	100	Fluidity	None
		Stain penetration	All
		Opacity	Transparent
		Color / oxidation	Dark brown (74)
			over stoneware
			Deep gold (17) over porcelain

SG 323 Medium Brown

Albany slip	90	Temperature	C/6–7
Cornwall stone	10	Surface @ C/6	Semigloss/high gloss
	100	Fluidity	None
		Stain penetration	Most
		Opacity	Translucent
		Color / oxidation	Medium brown
			(153)

SG 324 High-Gloss Medium Brown

Albany slip	90	Temperature	C/6
Nepheline syenite	10	Surface @ C/6	High gloss
	100	Fluidity	None
		Stain penetration	None
		Opacity	Opaque
		Color / oxidation	Medium brown
			(153)

SG 325 Semigloss Medium Brown

Albany slip	90	Temperature	C/6–8
Magnesium carbonate	10	Surface @ C/6	Semigloss/high gloss
	100	Fluidity	Little
		Stain penetration	None
		Opacity	Opaque
		Color / oxidation	Medium brown
			(153)

SG 326 Semigloss Army Green

Albany slip	90	Temperature	C/6–7
Talc	10	Surface @ C/6	Semigloss
	100	Fluidity	None
		Stain penetration	None
		Opacity	Opaque
		Color / oxidation	Army green (137)

SG 327 Greenish-brown

Albany slip	90	Temperature	C/6–8
Rutile	10	Surface @ C/6	Broken gloss/
	100		semigloss
		Fluidity	None
		Stain penetration	None
		Opacity	Opaque
		Color / oxidation	Warm greenish-brown (153)

SG 328 Gloss-to-Matt Brown

Albany slip	90	Temperature	C/6
Red iron oxide	10	Surface @ C/6	Gloss to matt
	100	Fluidity	None
		Stain penetration	Darks
		Opacity	Opaque
		Color / oxidation	Brown (73)

SG 329 Broken Gloss/Matt Brown

Albany slip	90	Temperature	C/6–8
Zinc oxide	10	Surface @ C/6	Broken gloss/matt
	100	Fluidity	Little
		Stain penetration	Darks
		Opacity	Opaque
		Color / oxidation	Medium brown (153)

SG 330 Gloss Medium Brown

Albany slip	90	Temperature	C/6–7
Soda ash	10	Surface @ C/6	Gloss
	100	Fluidity	Some
		Stain penetration	None
		Opacity	Opaque
		Color / oxidation	Medium brown (153)

SG 331 Opaque Medium Brown

Albany slip	90	Temperature	C/6–7
Spodumene	10	Surface @ C/6	High gloss
	100	Fluidity	None
		Stain penetration	Darks
		Opacity	Opaque
		Color / oxidation	Medium brown (153)

SG 332 Tan

Albany slip	80	Temperature	C/6–7
Rutile	20	Surface @ C/6	Smooth matt/semi-gloss
	100	Fluidity	None
		Stain penetration	Darks
		Opacity	Opaque
		Color / oxidation	Tan (35)

SG 333 High-Gloss Brown

Albany slip	80	Temperature	C/6–7
Magnesium carbonate	20	Surface @ C/6	High gloss
	100	Fluidity	Some
		Stain penetration	None
		Opacity	Opaque
		Color / oxidation	Brown (156)

SG 334 Broken-Gloss Brown

Albany slip	80	Temperature	C/6–7
Yellow ochre	20	Surface @ C/6	Broken gloss/semi-gloss
	100		
		Fluidity	Little
		Stain penetration	None
		Opacity	Opaque
		Color / oxidation	Brown (73)

SG 335 Translucent

Albany slip	80	Temperature	C/6–8
Zinc oxide	20	Surface @ C/6	High gloss
	100	Fluidity	None
		Stain penetration	Most
		Opacity	Translucent
		Color / oxidation	Tans (28) to medium brown (153)

SG 336 Pale Olive Green

Albany slip	80	Temperature	C/6–7
Whiting	20	Surface @ C/6	High gloss
	100	Fluidity	None
		Stain penetration	All
		Opacity	Transparent
		Color / oxidation	Pale olive green (159)

SG 337 Smooth-Matt Olive Green

Albany slip	80	Temperature	C/6–8
Titanium dioxide	20	Surface @ C/6	Smooth matt
	100	Fluidity	None
		Stain penetration	None
		Opacity	Opaque
		Color / oxidation	Olive green (147)

SG 338 Medium Brown

Albany slip	80	Temperature	C/6
Nepheline syenite	20	Surface @ C/6	Gloss/semigloss
	100	Fluidity	Little
		Stain penetration	Darks
		Opacity	Opaque when thick
			Translucent when thin
		Color / oxidation	Medium brown (153)

SG 339 Dark Olive Green

Albany slip	80	Temperature	C/6–8
Barium carbonate	20	Surface @ C/6	Satin matt
	100	Fluidity	None
		Stain penetration	Darks
		Opacity	Translucent
		Color / oxidation	Dark olive green (155)

SG 340 Dark Metallic Brown

Albany slip	80	Temperature	C/6–8
Burnt umber	20	Surface @ C/6	Smooth matt
	100	Fluidity	None
		Stain penetration	None
		Opacity	Opaque
		Color / oxidation	Dark metallic brown (73)

SG 341 Dark Brown

Albany slip	80	Temperature	C/6
Talc	20	Surface @ C/6	Gloss
	100	Fluidity	Some
		Stain penetration	None
		Opacity	Opaque
		Color / oxidation	Dark brown (73)

SG 342 Dark Brown

Albany slip	80	Temperature	C/6–8
Spodumene	20	Surface @ C/6	Gloss
	100	Fluidity	None
		Stain penetration	None
		Opacity	Opaque
		Color / oxidation	Dark brown (73)

SG 343 Brown

Albany slip	80	Temperature	C/6
Soda ash	20	Surface @ C/6	Gloss
	100	Fluidity	Little
		Stain penetration	Darks
		Opacity	Opaque
		Color / oxidation	Brown (154)

SG 344 Medium Brown/Army Green

Albany slip	80	Temperature	C/6–8
Zircopax	20	Surface @ C/6	Gloss
	100	Fluidity	None
		Stain penetration	None
		Opacity	Opaque
		Color / oxidation	Medium brown (156)/Army green (137)

SG 345 Brown/Olive Green

Albany slip	80	Temperature	C/6–7
Cryolite	20	Surface @ C/6	Gloss/semigloss
	100	Fluidity	None
		Stain penetration	Most
		Opacity	Translucent
		Color / oxidation	Brown/olive green (74/147)

SG 346 Mottled Blues

Albany slip	80	Temperature	C/6–8
Gerstley borate	20	Surface @ C/6	High gloss
	100	Fluidity	Little
		Stain penetration	Darks
		Opacity	Opaque
		Color / oxidation	Mottled—dark blue, light blue, and tan (90/87/160)

SG 347 Opaque Olive Green

Albany slip	80	Temperature	C/6–7
Wollastonite	20	Surface @ C/6	Gloss/semimatt
	100	Fluidity	Little
		Stain penetration	Darks
		Opacity	Opaque
		Color / oxidation	Olive green (138)

SG 348 Semimatt Charcoal

Albany slip	70	Temperature	C/6–8
Red iron oxide	30	Surface @ C/6	Semimatt
	100	Fluidity	None
		Stain penetration	None
		Opacity	Opaque
		Color / oxidation	Charcoal brown (123)

Note: Charcoal black (113) @ C/8

SG 349 Broken Army Green

Albany slip	70	Temperature	C/6–7
Talc	30	Surface @ C/6	Broken semigloss/ semimatt
	100		
		Fluidity	Little
		Stain penetration	Most
		Opacity	Translucent
		Color / oxidation	Army green (137)

SG 350 Satin-Matt Olive Green

Albany slip	70	Temperature	C/6
Whiting	30	Surface @ C/6	Satin matt
	100	Fluidity	Little
		Stain penetration	Most
		Opacity	Translucent when thick
			Clear when thin
		Color / oxidation	Olive green (147)

SG 351 Translucent

Albany slip	70	Temperature	C/6
Gerstley borate	30	Surface @ C/6	Gloss
	100	Fluidity	Little
		Stain penetration	Most
		Opacity	Translucent
		Color / oxidation	Blue/light tan
			(101/40)

SG 352 Matt Tan

Albany slip	70	Temperature	C/6
Titanium dioxide	30	Surface @ C/6	Matt
	100	Fluidity	Little
		Stain penetration	Darks
		Opacity	Opaque
		Color / oxidation	Tan (35)

SG 353 Charcoal Black

Albany slip	70	Temperature	C/6
Yellow ochre	30	Surface @ C/6	Semigloss/smooth
	100		matt
		Fluidity	Some
		Stain penetration	None
		Opacity	Opaque
		Color / oxidation	Charcoal black
			(113)

SG 354 Transparent Brown

Albany slip	70	Temperature	C/6
Soda ash	30	Surface @ C/6	Satin matt
	100	Fluidity	Little
		Stain penetration	All
		Opacity	Transparent
		Color / oxidation	Brown (73)

SG 355 Very Olive Green

Albany slip	70	Temperature	C/6–7
Nepheline syenite	30	Surface @ C/6	High gloss
	100	Fluidity	Little
		Stain penetration	None
		Opacity	Opaque
			Translucent when very thin
		Color / oxidation	Very olive green (147)

SG 356 Semigloss/Matt Tan

Albany slip	70	Temperature	C/6–8
Zircopax	30	Surface @ C/6	Semigloss/matt
	100	Fluidity	None
		Stain penetration	Darks
		Opacity	Opaque
		Color / oxidation	Tan (28)

SG 357 Translucent Olive Green

Albany slip	70	Temperature	C/6
Wollastonite	30	Surface @ C/6	Broken gloss/matt
	100	Fluidity	Some
		Stain penetration	Most
		Opacity	Translucent
			Transparent when very thin
		Color / oxidation	Olive green (137)

SG 358 Transparent Olive Green Dark

Albany slip	70	Temperature	C/6–8
Barium carbonate	30	Surface @ C/6	High gloss
	100	Fluidity	None
		Stain penetration	All
		Opacity	Transparent
		Color / oxidation	Dark olive green (137)

SG 359 Dark Greenish-Brown

Albany slip	70	Temperature	C/6
Cornwall stone	30	Surface @ C/6	Broken gloss/matt
	100	Fluidity	Some
		Stain penetration	Darks
		Opacity	Translucent
		Color / oxidation	Dark greenish-brown (155)

SG 360 Very Dark Marine Blue

Albany slip	51	Temperature	C/6–8
Buckingham feldspar	35	Surface @ C/6	High gloss
Kentucky ball	8	Fluidity	None
Whiting	4	Stain penetration	None
Cobalt oxide	2	Opacity	Opaque
	100	Color / oxidation	Very dark marine blue (90)

SG 361 Broken Cool Browns

Albany slip	50	Temperature	C/6–8
Spodumene	50	Surface @ C/6	High gloss
	100	Fluidity	None
		Stain penetration	Most
		Opacity	Translucent
		Color / oxidation	Broken cool browns (73)

SG 362 Opaque Brown

Albany slip	50	Temperature	C/6–8
Redart	50	Surface @ C/6	Semigloss
	100	Fluidity	None
		Stain penetration	None
		Opacity	Opaque
		Color / oxidation	Brown (154)

SG 363 Medium Brown Slip

Albany slip	50	Temperature	C/6–7
Nepheline syenite	48	Surface @ C/6	High gloss
Chrome oxide	2	Fluidity	None
	100	Stain penetration	None
		Opacity	Opaque
		Color / oxidation	Medium brown (153)

SG 364 Hazy Bluish-Brown

Albany slip	50	Temperature	C/6
Magnesium carbonate	40	Surface @ C/6	Broken semigloss/
Whiting	10		semimatt
	100	Fluidity	Some
		Stain penetration	None
		Opacity	Opaque
		Color / oxidation	Hazy bluish-brown (90/73)

SG 365 Broken Opaque Medium Brown

Albany slip	50	Temperature	C/6–7
Cornwall stone	40	Surface @ C/6	Broken gloss/matt
Rutile	8	Fluidity	Little
Red copper oxide	2	Stain penetration	Darks
	100	Opacity	Opaque
		Color / oxidation	Medium brown (153)

SG 366 Transparent Medium Brown

Albany slip	50	Temperature	C/6–7
Colemanite	30	Surface @ C/6	Gloss
Spodumene	10	Fluidity	Little
Zinc oxide	5	Stain penetration	All
Strontium carbonate	5	Opacity	Transparent
	100	Color / oxidation	Medium brown (153)

SG 367 Blackish Brown

Albany slip	50	Temperature	C/6–7
Whiting	25	Surface @ C/6	Satin matt
Burnt umber	25	Fluidity	Some
	100	Stain penetration	None
		Opacity	Opaque
		Color / oxidation	Dark blackish-brown (113/73)

SG 368 Light Blue-Gray

Tennessee ball #1	50	Temperature	C/6–8
Sodium bicarbonate	25	Surface @ C/6	Gloss
Zinc oxide	18	Fluidity	None
Whiting	7	Stain penetration	Most
	100	Opacity	Translucent
		Color / oxidation	Light blue-gray (104)

SG 369 Opaque Dark Brown

Albany slip	50	Temperature	C/6–8
Spodumene	25	Surface @ C/6	Gloss
Soda ash	15	Fluidity	None
Yellow ochre	10	Stain penetration	None
	100	Opacity	Opaque
		Color / oxidation	Dark brown (73)

SG 370 Speckled Blue-Gray

Albany slip	45	Temperature	C/6–8
Potash feldspar	27	Surface @ C/6	High gloss
Flint	9	Fluidity	None
Zinc oxide	9	Stain penetration	None
Rutile	5	Opacity	Opaque
Whiting	5	Color / oxidation	Speckled blue-gray
	100		(96)

SG 371 Black-Brown

Albany slip	45	Temperature	C/6–8
Soda feldspar	25	Surface @ C/6	Gloss
Zinc oxide	20	Fluidity	Some
Red iron oxide	10	Stain penetration	None
	100	Opacity	Opaque
Ball clay	5	Color / oxidation	Black-brown

SG 372 Charcoal Black Satin Matt

Tennessee ball #1	44	Temperature	C/6–8
Barnard clay	44	Surface @ C/6	Satin matt
Whiting	12	Fluidity	None
	100	Stain penetration	None
		Opacity	Opaque
		Color / oxidation	Charcoal black
			(113)

SG 373 Broken Browns

Albany slip	42	Temperature	C/6
Zinc oxide	29	Surface @ C/6	Broken gloss/semi-gloss
Wollastonite	24		
Vanadium pentoxide	5	Fluidity	Some
	100	Stain penetration	Darks
		Opacity	Translucent
		Color / oxidation	Broken browns (73)

SG 374 Semigloss Browns

Albany slip	40	Temperature	C/6–8
Colemanite	40	Surface @ C/6	Semigloss
Yellow ochre	20	Fluidity	None
	100	Stain penetration	None
		Opacity	Opaque
		Color / oxidation	Browns (154)

SG 375 Gray

Albany slip	40	Temperature	C/6–8
Gerstley borate	30	Surface @ C/6	Semimatt
Flint	27	Fluidity	None
Zircopax	3	Stain penetration	None
	100	Opacity	Opaque
		Color / oxidation	Gray (8)

CONE/8

SG 376 Oil Puddle Blue-Gray

Albany slip	70	Temperature	C/8–9
Burnt umber	30	Surface @ C/8	Semigloss/smooth matt
	100	Fluidity	None
		Stain penetration	None
		Opacity	Opaque
		Color / oxidation	Oil puddle blue-gray (90–113)

SG 377 Warm Light Brown

Albany slip	70	Temperature	C/8
Rutile	30	Surface @ C/8	Matt
	100	Fluidity	Some
		Stain penetration	None
		Opacity	Opaque
		Color / oxidation	Warm light brown (36)

SG 378 Greenish-Brown

Albany slip	60	Temperature	C/8–9
Whiting	40	Surface @ C/8	Gloss/smooth matt
	100	Fluidity	None
		Stain penetration	Darks
		Opacity	Opaque
		Color / oxidation	Greenish-brown (155)

SG 379 Light Tan

Albany slip	50	Temperature	C/8–9
Rutile	40	Surface @ C/8	Semimatt
Gerstley borate	10	Fluidity	Little
	100	Stain penetration	Darks
		Opacity	Opaque
		Color / oxidation	Light tan (29)

SG 380 Speckled Light Tan

Albany slip	50	Temperature	C/8–9
Talc	40	Surface @ C/8	Smooth matt
Whiting	10	Fluidity	None
	100	Stain penetration	None
		Opacity	Opaque
		Color / oxidation	Speckled light tan (39)

SG 381 Gray-Tan

Albany slip	50	Temperature	C/8
Soda (Del Monte) spar	25	Surface @ C/8	Gloss
Zircopax	25	Fluidity	None
	100	Stain penetration	Darks
		Opacity	Opaque
		Color / oxidation	Gray-tan with brown speckles (39)

SG 382 Light and Medium Olives

Albany slip	36	Temperature	C/8–9
Flint	30	Surface @ C/8	Smooth matt
Whiting	17	Fluidity	Little
Kaolin	10	Stain penetration	None
Lithium carbonate	5	Opacity	Opaque
Chrome oxide	2	Color / oxidation	Broken light and
	100		medium olives (143/138)

SG 383 Light Grayish-White

Kaolin	30	Temperature	C/8–9
Kona #4 spar	25	Surface @ C/8	Dry matt
Whiting	25	Fluidity	None
Kentucky ball #4	20	Stain penetration	All
	100	Opacity	Transparent
		Color / oxidation	Light grayish-white

CONE/9

SG 384 Transparent Matt

Kentucky ball	51	Temperature	C/9
Gerstley borate	30	Surface @ C/9	Matt
Barium carbonate	10	Fluidity	None
Lithium carbonate	9	Stain penetration	All
	100	Opacity	Transparent
		Color / oxidation	Porcelain: tan (40)
			Stoneware: brown (73)

SG 385 Broken Tan/Brown

Ball clay	45	Temperature	C/9–10
Barium carbonate	25	Surface @ C/9	Matt
Gerstley borate	20	Fluidity	None
Whiting	10	Stain penetration	Darks
	100	Opacity	Opaque
		Color / oxidation	Broken tans and
			browns (40/73)

SG 386 Blue Speckles

Albany slip	40.0	Temperature	C/9–10
Ball clay	24.0	Surface @ C/9	Gloss
Wollastonite	19.0	Fluidity	None
Nepheline syenite	12.5	Stain penetration	None
Rutile	4.0	Opacity	Opaque
Cobalt carbonate	0.5	Color / oxidation	Grayish-green with
	100.0		blue speckles (140)

SG 387 Clear Tan

Talc	31	Temperature	C/9–10
Georgia kaolin	30	Surface @ C/9	Gloss
Kentucky ball #4	21	Fluidity	None
Frit #54	10	Stain penetration	Most
Lithium carbonate	8	Opacity	Translucent
	100	Color / oxidation	Clear tan over
			off-white (40/30)

CONE/10

SG 388 Pale Yellowish-Gray

Albany slip	40	Temperature	C/10–11
Lepidolite	35	Surface @ C/10	Semigloss
Whiting	13	Fluidity	None
Flint	12	Stain penetration	Most
	100	Opacity	Translucent
		Color / oxidation	Speckled pale
			yellowish-gray (23)

VI

Stoneware Glazes

The history of ceramics begins over 8,000 years ago, in the Neolithic period, with the development of the earliest human civilizations. The clay used by the first potters was buff, gray, or light red earthenware; the pottery they produced was semisoft, porous (not waterproof), and sometimes burnished. Using colored engobes and textural designs, the potters of the early civilizations—Greek; Roman; Mesopotamian; Egyptian; Chinese; Korean; Japanese; and North, Central, and South American—decorated their wares with geometric patterns and stylized animal and human forms. Although the creation of ceramics was well established throughout the ancient world, the use of vitreous, or glass-like—and waterproof—glazes was not. By 3000 B.C. only Egypt and Mesopotamia had developed lead and tin glazes. The discovery of these glazes significantly increased the scope of available colors and surface effects and, as a result, extended the range of ceramic use.

The development of stoneware occurred at about the same time. As early as 4,000 years ago, in China, stoneware was being fired in wood-burning kilns. During the firing, as was noted on page 122, ash from the kiln fire settled on the shoulders and other exposed areas of the stoneware, creating a natural glaze. Stoneware glazes have been used in neighboring Korea for over 2,000 years; while in Japan glazed stoneware dates only from the thirteenth century,

148

just a little over 700 years ago. The production of glazed stoneware was not introduced into Europe until the sixteenth century, when potters in Germany first attempted to copy the red stoneware imported from China as part of the tea trade. Later still, in the seventeenth century, the first salt-glazed white stoneware was produced in England as a substitute for the rare and expensive Chinese porcelain. Even today stoneware continues to be a popular medium, especially for the artist-potter. Stoneware, which is generally fired in the C/4 to C/12 range, midway between earthenware and porcelain firing temperatures, is very hard, "rings" when struck, and can be either hand-built or thrown on a potter's wheel.

In working with stoneware glazes, potters who use electric kilns may face a problem. An electric kiln designed to fire up through C/6 can reach C/8, but the life of its heating elements will be shortened as a result.* Cone/10 bodies—like stoneware—can be fired at C/6 temperatures, but the results will never be as satisfactory as with clay bodies specifically compounded for C/6 temperatures; in particular, the bodies will not vitrify. (To enable higher-range bodies to vitrify at C/6 temperatures, an additional 2 to 15 percent of flux or earthenware clay should be wedged into the prepared clay to lower the vitrification point.) Wares glaze-fired at lower-end stoneware temperatures, particularly in the C/6 range, will take on the iron spotting that is characteristic of stoneware, but the spotting will be less prominent than at higher temperatures (generally C/9 to C/10), because the iron in the clay will not readily bleed through the glaze. And since, like other bodies fired at lower-than-recommended temperatures, stoneware-glaze pieces will not vitrify, the body and glaze will not form the integrated layer that gives stoneware its other distinctive features: soft, muted colors and hard surface.

The potter's dilemma can be solved if he or she uses clay bodies that mature at C/6 and selects glazes that do not develop the "pasty" look that lower-heat electric firing often produces. In preparing clay and glaze, the ceramist should adopt one or more of the stoneware-glaze techniques outlined below. Whether electric- or gas-fired, the

---

* The reverse is true for a kiln designed to fire up through the C/10 range which is fired up through C/8 only; the life expectancy of its heating elements, and of the kiln itself, will be extended. Furthermore, at the lower temperature the firings will cost less and will take less time, and shrinking and warping will both be reduced.

glazes should take on a reduced-stoneware appearance (see below, pages 151–54, for an explanation of reduction firing.)

1. Blending additional granulated iron, ilmenite, manganese dioxide, iron chromate, or screened iron filings into the clay body will produce the characteristic iron spots, which will bleed through the glaze.

2. Introducing the same metallic grains into the glaze in amounts of 0.1 to 1 percent will produce speckling in the glaze. Because of their weight, however, the metallic grains have a tendency to settle to the bottom of the glaze. This tendency can be alleviated either by decanting some of the water from the glaze bucket or by adding bentonite, CMC (or other glues), ball clay, or starch to thicken the glaze enough to keep the particles in suspension.

3. Altering the glaze formula to include lithium will help draw the iron from the clay body to the glaze surface and thus increase the amount of spotting. Lithium can be introduced either as spodumene, lepidolite, or petalite feldspars which contain lithium; or as lithium carbonate. Unfortunately, all of these minerals are expensive to use. Moreover, because lithium is an extremely active flux, its replacement or addition must be carefully tested in a formula. In some cases the addition of 1 to 5 percent molybdenum or titanium will produce results similar to those of lithium.

4. The addition of 3 to 7 percent of either rutile, titanium, molybdenum, or zinc in glazes will prevent the glaze from acquiring the "pasty" look referred to above and will create, instead, an intriguing pattern of streaks and mottled or broken colors. When used in amounts of 5 to 12 percent, though, these same minerals may allow crystalline matts, fluid streaks, or bubbling (in thick areas) to develop in the glaze. Furthermore, in certain glazes, these minerals become active fluxes that cause the glaze to turn fluid.

5. *Dual-glazing,* the application of one glaze over another glaze, can produce a variety of effects on the stoneware surface. In dual-glazing, the under-, or base, glaze will bleed, spot, show through, or in some other way influence the over-glaze. When a fluid (glossy), dark base glaze is covered by a light-colored, matt, nonflowing over-glaze, the under-glaze will break through the over-glaze to give a smoothed-over-craters, streaked, lizard-skin, or gloss-matt appearance. Although the colors can be reversed, the glaze order cannot.

Also, a matt base glaze covered with a fluid over-glaze does not produce a desirable effect.

6. If a very dark clay body (or a clay body treated with a very dark engobe) is covered with a translucent pastel over-glaze, the darker body, which will show through the glaze, will provide a striking contrast to the glaze. Other methods of glazing that allow the body to show through are wax resist and partial glazing; or the ceramist can simply wipe away some of the glaze.

Reduction Firing of Stoneware Glazes

Another way for ceramists to achieve iron spotting, mottled pattern, and soft, muted colors is to rely on reduction firing. It is possible to obtain the desired effect in electric kilns; the finished pots will be similar to those produced in oil-, gas-, or wood-fueled kilns. An awareness of the reduction-firing process itself will help the ceramist understand why glaze effects may vary.

In reduction firing, all metals combine, at the outset, with oxygen to form metallic oxides. The proportion of oxygen to metal varies; for example, FeO (ferrous oxide) contains proportionately more iron than Fe_2O_3 (ferric oxide) or $FeCO_3$ (ferrous carbonate). If the amount of oxygen in Fe_2O_3 and $FeCO_3$ is reduced, the two compounds will both become FeO; if all the oxygen is taken away, just the metal (Fe) will be left. The process of oxygen removal is known as *reduction*. (The reverse process, called *oxidation,* is the addition of oxygen to FeO to form ferric oxide and ferrous carbonate.) In glaze firing, the metal and its oxygen are held together by their affinity, or attractive force. If carbon monoxide (CO) is introduced into the glaze at temperatures above 1000°F, the carbon becomes "hungry" for oxygen and quickly converts itself to carbon dioxide (CO_2). The carbon seizes oxygen from whatever source it can—air, glaze and clay body minerals, and metallic oxides. It will, in fact, take as much of the available oxygen as possible from any source to become carbon dioxide. To activate the reduction process, then, the ceramist must obtain a carbon monoxide atmosphere in the kiln. There are several ways of doing so:

1. The firing of kilns with either oil, natural gas, manufactured gas, kerosene, wood, or propane fuels is the easiest method. The

amount of air entering the firebox is reduced to allow smoke, which consists of minute particles of carbon, to enter the kiln. The limited amount of oxygen in the air permits the particles of carbon to form carbon monoxide and hydrocarbon gas; this process creates the reduction atmosphere.

2. Artificial reduction is produced by the addition of silicon carbonate (400 to 600 mesh) to the glaze in amounts of 0.1 to 0.5 percent. When the glaze is fired at higher temperatures, the carbon becomes active, combining with the oxygen in the glaze to become carbon dioxide, which then dissipates as a gas. The addition of too much carbonate, or the use of a coarse-grained carbonate, will produce pits and bubbles in the glaze surface. Fine-grained carbonate in small amounts, on the other hand, escapes through the glaze as a gas, creating only small bubbles which are relatively easy to heal in a glossy, slightly fluid glaze.

3. In an electric kiln, the introduction of organic matter like thin strips of wood, mothballs, charcoal, rags, or rolled-up newspaper into the lowest spy hole is the reduction technique used most often. The carbon from the organic material combines with the limited supply of oxygen in the kiln and in the ceramic pieces to form the carbon monoxide atmosphere; the process is very much like that in a gas kiln. While frequent reduction firing in an electric kiln will shorten the life of the elements from 10 to 20 percent, moderate use will not have a depreciating effect.

Reduction Firing Techniques

Reduction firing is not difficult. The procedure that follows applies primarily to gas firing, but it will work for wood, coal, oil, or even electric firing as well. The glazeware is first loaded into the kiln, and the pilot light or lowest setting of the burners is turned on. This step is called *candling* and is done to preheat the kiln and to remove moisture from the kiln, the clay bodies, and the glazes. The temperature is then brought up slowly, until it reaches 600°F, after which the kiln is fired as fast as the wares, shelves, and burners can tolerate without breaking. At 1600° to 1800°F the *first* reduction, called *body reduction,* occurs. In the case of a fuel kiln (any but an electric kiln), the dampers are closed down, though not necessarily closed completely, and the primary and secondary air is reduced. Which step is taken depends upon the kiln; in some kilns it is sufficient to dampen the flue, while in others the air blower will also have to be turned off, with primary and secondary air sources controlled. When a wisp of smoke comes out of the bottom-most kiln spy hole, the settings for dampers and air are correct. Although different kiln designs will require different adjustments, the basic concept is the same—less air and a wisp of smoke coming out of the bottom spy hole. For most reduction firing no more smoke than a wisp is needed. Reduction in an electric kiln includes the introduction of organic matter into the bottom-most spy hole. The kiln should be stacked in such a way as to leave a clear space in front of the spy hole so that the organic matter will not drop onto the pots. From fifteen to thirty minutes of reduction is adequate for the average-size kiln, regardless of the fuel used or of kiln design. However, a heavy reduction in a large kiln may take as long as forty-five minutes.

When the reduction is finished, the dampers are opened and the firing process is continued with a slightly reductive or a neutral atmosphere. Often during the body reduction phase, the temperatures inside the kiln will level off and increase very little. After the kiln has reached the maturing temperature, the firing cone has bent, and the top and bottom temperatures have been balanced, the *second* reduction, or *glaze reduction,* is begun. The technique followed and the length of time allowed are the same as for body reduction. During reduction the temperature may drop about 100°, because of the cooler flame.

After reduction, many potters run a neutral or oxidative firing for an additional five minutes. This short firing will clear out the carbon in the kiln and flash the surface of the clay and glaze with oxygen, thus converting the surface iron from ferrous oxide back to ferric oxide. The conversion will cause the cool iron colors of the clay and glaze to "warm up," but the firing is too short to alter the copper reds and celadons appreciably. To illustrate color differences resulting from reduction and oxidation firings, if two glazed pots with iron in the celadon glaze and in the clay body are fired in two different atmospheres—one oxidative and one reductive—the reduced pot will have a cool clay body and a blue-green celadon glaze, while the other will have a warm body and a green-jade glaze. During reduction, the iron turned into ferrous oxide (a cool color), while the oxidation firing turned the iron into yellow ferric oxide (a warm color). It was the firing of English china in saggers to protect it from the slightly reductive atmosphere that originally gave that porcelain its creamy color. Because true Chinese porcelains, on the other hand, were reduced, they took on a cool white color.

STONEWARE GLAZE FORMULAS

The glaze formulas in this section are for C/4, C/5, C/6, and C/8 firings with gloss, semigloss, semimatt, matt, and dry matt finishes in addition to their transparent, translucent, or opaque qualities. Most formulas list both oxidation and reduction color results, to facilitate the comparison of any differences. With some formulas, particularly for the clear and translucent glazes, the two firing methods produce roughly the same color or colors.

CONE/4

G 3026 Transparent Clear

Lepidolite	80	Temperature	C/4–6
Gerstley borate	17	Surface @ C/4	Satin matt
Lithium carbonate	3	Fluidity	None
	100	Stain penetration	All
		Opacity	Transparent
Bentonite	2	Color / oxidation	Clear
		Note: Little cracks	
Uranium	10	Color / oxidation	Clear light tan (40)
Vanadium pentoxide stain	7	Color / oxidation	Clear light green (136)

G 3027 Opaque

Oxford feldspar	51.0	Temperature	C/4–5
Flint	24.0	Surface @ C/4	Gloss
Whiting	13.0	Fluidity	Little
E.P. kaolin	3.8	Stain penetration	Darks
Lithium carbonate	3.5	Opacity	Opaque
Magnesium carbonate	2.7	Color / oxidation	Light pinkish-
Strontium carbonate	2.0		purple (36)
	100.0		

G 3028 Semigloss

Oxford feldspar	43.2	Temperature	C/4–5
Flint	24.0	Surface @ C/4	Semigloss
Colemanite	20.6	Fluidity	None
Barium carbonate	6.4	Stain penetration	All
Zinc oxide	3.2	Opacity	Translucent
Whiting	1.5	Color / oxidation	Clear
China clay	1.1		
	100.0		

G 3029 Alfred University Dry Matt

Flint	38.7	Temperature	C/4–6
Nepheline syenite	31.2	Surface @ C/4	Dry matt
Dolomite	17.6	Fluidity	Some flow
Georgia kaolin	6.4	Stain penetration	Darks
Whiting	3.4	Opacity	Opaque
Zinc oxide	2.7	Color / oxidation	White (7)
	100.0		

G 3030 Clear

Spodumene	21	Temperature	C/4–5
Frit #14 HOMMEL	18	Surface @ C/4	Semigloss
Flint	18	Fluidity	None
Barium carbonate	16	Stain penetration	All
Lepidolite	13	Opacity	Transparent
Georgia china clay	11	Color / oxidation	Clear
Whiting	3		
	100		

CONE/5

G 3031 White

Del Monte feldspar	89	Temperature	C/5–6
Barium carbonate	6	Surface @ C/5	Gloss
Whiting	5	Fluidity	None
	100	Stain penetration	All
		Opacity	Transparent
		Color / oxidation	White (32)
		/ reduction	Cool white (7)
Rutile	10	Surface @ C/5	Semimatt
		Fluidity	Little
		Stain penetration	None
		Opacity	Opaque
		Color / oxidation	Tan (24)
		/ reduction	Dark green (129)

G 3032 Translucent White

Nepheline syenite	44	Temperature	C/5–6
E.P. kaolin	19	Surface @ C/5	Semigloss
Whiting	17	Fluidity	Little
Zinc oxide	15	Stain penetration	Most
Flint	5	Opacity	Translucent
	100	Color / oxidation	White (32)
		/ reduction	White (32)

G 3033 Tans

Frit #14 HOMMEL	40	Temperature	C/5
Whiting	30	Surface @ C/5	Semigloss
Nepheline syenite	28	Fluidity	Some
Bentonite	2	Stain penetration	Most
	100	Opacity	Translucent
		Color / reduction	Tans (28)
Molybdenum	8	Stain penetration	Darks
		Opacity	Opaque
		Color / reduction	Tans (28/38)

Selenium	10 ⎫	Color / reduction	Tan (28)
Tin oxide	4 ⎭		

CONE/6

G 3034 Simpson's Grayish-White

Nepheline syenite	75.5	Temperature	C/6–7
Whiting	9.4	Surface @ C/6	Gloss
E.P. kaolin	9.4	Fluidity	Little
Barium carbonate	3.8	Stain penetration	All
Bone ash	1.9	Opacity	Transparent
	100.0	Color / oxidation	White clear
		/ reduction	Grayish-white clear
			(7)

G 3035 Petalite White

Petalite	70	Temperature	C/6–8
Talc	14	Surface @ C/6	
Whiting	11	/ oxidation	Satin matt
E.P. kaolin	5	/ reduction	Gloss
	100	Fluidity	
		/ oxidation	Little
		/ reduction	Little
		Stain penetration	
		/ oxidation	All
		/ reduction	Darks
		Opacity	
		/ oxidation	Transparent
		/ reduction	Opaque
		Color / oxidation	Clear
		/ reduction	White (32)

G 3036 Oxford White

Oxford feldspar	67	Temperature	C/6–8
Whiting	20	Surface @ C/6	Semigloss
Flint	8	Fluidity	None
E.P. kaolin	5	Stain penetration	All
	100	Opacity	Translucent
		Color / reduction	White (32)

G 3037 Newcomb Gloss

Frit #3304 FERRO	63.9	Temperature	C/6–8
Oxford feldspar	22.2	Surface @ C/6	Gloss
Whiting	6.6	Fluidity	None
Flint	3.8	Stain penetration	All
Zinc oxide	3.5	Opacity	Transparent
	100.0	Color / oxidation	Clear
		/ reduction	Clear

G 3038 White Gloss

Spodumene	50	Temperature	C/6
Talc	41	Surface @ C/6	Gloss
Frit #14 HOMMEL	7	Fluidity	Little
Bentonite	2	Stain penetration	Most
	100	Opacity	Opaque
		Color / oxidation	White (31)
		/ reduction	Clear white (15)

G 3039 Feldspar White

Feldspar F-4	46.6	Temperature	C/6–8
Barium carbonate	23.5	Surface @ C/6	
Tennessee ball #1	11.1	/ oxidation	Satin matt
Whiting	9.8	/ reduction	Dry matt
Zinc oxide	9.0	Fluidity	Little
	100.0	Stain penetration	All
		Opacity	Translucent
		Color / oxidation	White (7)
		/ reduction	White (7)

G 3040 Semigloss White

Plastic vitrox	44	Temperature	C/6–8
Colemanite	44	Surface @ C/6	Semigloss
Zircopax	12	Fluidity	Little
	100	Stain penetration	All
		Opacity / oxidation	Translucent
		/ reduction	Transparent
		Color / oxidation	White (32)
		/ reduction	White (32)

G 3041 Transparent

Kingman feldspar	43.3	Temperature	C/6–8
Whiting	17.6	Surface @ C/6	Gloss
Flint	10.7	Fluidity	None
Nepheline syenite	9.1	Stain penetration	All
Colemanite	8.3	Opacity	Transparent
E.P. kaolin	8.1	Color / oxidation	Clear slight green
Zinc oxide	2.9		(136)
	100.0		

| Rutile | 3.0 } | Color / oxidation | Clear, light chrome |
| Copper carbonate | 1.0 } | | green; yellow; and tans (125/3/29) |

| Rutile | 3.0 } | Color / oxidation | Clear |
| Titanium | 1.0 } | | |

G 3042 Clear

Oxford feldspar	37.4	Temperature	C/6–7
Barium carbonate	24.7	Surface @ C/6	Semigloss/semimatt
Colemanite	22.9	Fluidity	Little
Whiting	11.1	Stain penetration	All
Magnesium carbonate	3.9	Opacity	Transparent
	100.0	Color / oxidation	Clear
		/ reduction	Clear

G 3043 Dark Brown

Hardwood ash, mixed	35	Temperature	C/6–8
Yellow ochre	35	Surface @ C/6	Broken matt/gloss
Whiting	30	Fluidity	Little
	100	Stain penetration	None
		Opacity	Opaque
		Color / oxidation	Dark brown (73)
		/ reduction	Dark brown with some shine (73)

G 3044 Barium Matt

Barium carbonate	34	Temperature	C/6–8
Custer feldspar	27	Surface @ C/6	Semigloss/matt
Flint	21		volcanic
Georgia kaolin	9	Fluidity	Little
Zinc oxide	9	Stain penetration	Most
	100	Opacity	Translucent
		Color / oxidation	White (32)
		/ reduction	White (7)

G 3045 Light Tan

Whiting	33.3	Temperature	C/6–8
Pumice (volcanic ash)	33.3	Surface @ C/6	Broken gloss/matt
Colemanite	33.3	Fluidity	Little
	99.9	Stain penetration	Most
		Opacity	Translucent
		Color / reduction	Light tan with greenish tones (29/157)

G 3046 Pinks

Cornwall stone	27.5	Temperature	C/6–8
Georgia kaolin	23.8	Surface @ C/6	Gloss
Frit #3110 FERRO	18.2	Fluidity	None
Whiting	13.7	Stain penetration	All
Flint	9.1	Opacity	Transparent
Lithium carbonate	7.7	Color / oxidation	Clear, light pinkish-gray (45/152)
	100.0		
		/ reduction	Clear, light pinkish-gray with deeper pink

G 3047 Sheri "G-4"

Nepheline syenite	24.4	Temperature	C/6–8
Flint	20.1	Surface @ C/6	Gloss/broken matt
Georgia kaolin	18.2	Fluidity	Little
Whiting	17.6	Stain penetration	All
Colemanite	10.9	Opacity	Transparent
Talc	8.8	Color / oxidation	Clear
	100.0	/ reduction	Clear

CONE/8

G 3048 Kingman Opaque Gloss

Kingman feldspar	71.4	Temperature	C/8–9
Cornwall stone	17.2	Surface @ C/8	Gloss
Zinc	5.7	Fluidity	Little
E.P. kaolin	5.7	Stain penetration	Darks
	100.0	Opacity	Opaque
		Color / oxidation	White (32)
		/ reduction	Blue-gray (100)

G 3049 High Gloss

Nepheline syenite	65.1	Temperature	C/8–9
Frit #25	13.0	Surface @ C/8	High gloss
Whiting	6.5	Fluidity	Little
Vanadium	7.0	Stain penetration	
Georgia kaolin	3.7	/ oxidation	All
Lithium carbonate	2.8	/ reduction	None
Zinc oxide	1.9	Opacity / oxidation	Transparent
	100.0	/ reduction	Opaque
		Color / oxidation	Clear
		/ reduction	Blue-black (113/90)

G 3050 Nancy's 3-D

Nepheline syenite	62.4	Temperature	C/8–10
Dolomite	20.8	Surface @ C/8	Gloss
Tin oxide	9.8	Fluidity	None
Tennessee ball #4	4.1	Stain penetration	None
Bentonite	2.9	Opacity	Opaque
	100.0	Color / oxidation	White (32)
		/ reduction	Blue-white (7)

G 3051 Keystone

Keystone feldspar	59.7	Temperature	C/8–9
Whiting	17.9	Surface @ C/8	
E.P. kaolin	15.9	/ oxidation	Semimatt
Rutile	4.0	/ reduction	Semigloss
Red iron oxide	2.0	Fluidity	None
Bentonite	.5	Stain penetration	Darks
	100.0	Opacity	Opaque
		Color / oxidation	Brown (156)
		/ reduction	Light tan (160)

G 3052 Sage Marshmallow

Oxford soda feldspar	57.6	Temperature	C/8–9
Whiting	17.0	Surface @ C/8	Matt
Georgia kaolin	15.0	Fluidity	None
Rutile	4.8	Stain penetration	Most
Zinc oxide	3.7	Opacity	Translucent
Flint	1.9	Color / oxidation	Tan (38)
	100.0	/ reduction	Blue-gray (9/100)

G 3053 Slater's Tan

Custer feldspar	48.3	Temperature	C/8–9
Whiting	17.5	Surface @ C/8	
E.P. kaolin	14.4	/ oxidation	Satin matt
Rutile	8.8	/ reduction	Semigloss
Flint	4.8	Fluidity	Little
Iron oxide	4.4	Stain penetration	Darks
Zinc oxide	1.8	Opacity	Opaque
	100.0	Color / oxidation	Tan (27)
		/ reduction	Tan (35)

G 3054 Gloss White

Nepheline syenite	47.9	Temperature	C/8–9
Flint	21.3	Surface @ C/8	Gloss
Whiting	19.5	Fluidity	None
E.P. kaolin	8.8	Stain penetration	All
Zinc oxide	2.5	Opacity	Transparent
	100.0	Color / oxidation	White, clear (32)
		/ reduction	Blue-white (7)

G 3055 Buttery

Oxford feldspar	47.3	Temperature	C/8
Flint	29.0	Surface @ C/8	Gloss
Whiting	10.0	Fluidity	Some
Dolomite	6.2	Stain penetration	
Zinc oxide	5.8	/ oxidation	Darks
Georgia kaolin	1.7	/ reduction	None
	100.0	Opacity / oxidation	Opaque
		/ reduction	Opaque
		Color / oxidation	White (32)
		/ reduction	Cool-white (7)

G 3056 Translucent Matt

Kingman feldspar	46.2	Temperature	C/8–10
Barium carbonate	23.6	Surface @ C/8	Matt
Kentucky ball #4	11.2	Fluidity	None
Whiting	10.0	Stain penetration	Most
Zinc oxide	9.0	Opacity	Translucent
	100.0	Color / oxidation	White (32)
		/ reduction	Cool-white (104)

G 3057 Tapeazo Translucent

Nepheline syenite	44.1	Temperature	C/8
Whiting	39.2	Surface @ C/8	Semigloss
Feldspar F-4	9.8	Fluidity	Little
Silica	4.9	Stain penetration	All
Bentonite	2.0	Opacity	Translucent
	100.0	Color / oxidation	Light tan (40)
		/ reduction	Light greenish-gray (134)

G 3058 Lepidolite

Lepidolite	44	Temperature	C/8
Flint	35	Surface @ C/8	Gloss
Whiting	15	Fluidity	None
Zinc oxide	6	Stain penetration	All
	100	Opacity	Transparent
		Color / oxidation	Clear with white specks
		/ reduction	Grayish blue-white (7)

G 3059 PV White (Long Beach State)

Plastic vitrox	43.5	Temperature	C/8
Colemanite	43.5	Surface @ C/8	Gloss
Zircopax	13.0	Fluidity	Little
	100.0	Stain penetration	Most
		Opacity	Translucent
		Color / oxidation	White (32)
		/ reduction	Cool white (7)

G 3060 Opaque White

Keystone feldspar	40	Temperature	C/8–9
Flint	37	Surface @ C/8	Semigloss when thick
Whiting	17		Dry matt when thin
Talc	6	Fluidity	Some
	100	Stain penetration	None
		Opacity	Opaque
		Color / oxidation	White (32)

G 3061 Power Blue

Keystone feldspar	36.4	Temperature	C/8–9
Flint	23.7	Surface @ C/8	Gloss
Whiting	10.1	Fluidity	Little
Tin oxide	9.8	Stain penetration	None
Barium carbonate	9.3	Opacity	Opaque
Georgia kaolin	6.4	Color / oxidation	Power blue (88)
Zinc oxide	2.5		
Dolomite	1.6		
Cobalt carbonate	0.2		
	100.0		

G 3062 Smooth Matt

Talc	33	Temperature	C/8–10
Nepheline syenite	23	Surface @ C/8	Smooth matt
E.P. kaolin	18	Fluidity	Little
Whiting	15	Stain penetration	Darks
Titanium dioxide	11	Opacity	Opaque
	100	Color / oxidation	White (31)
		/ reduction	Blue-gray (100)

G 3063 Magnesia

Kingman feldspar	31.2	Temperature	C/8–9
Talc	22.3	Surface @ C/8	Gloss
Flint	17.9	Fluidity	None
E.P. kaolin	17.9	Stain penetration	Most
Gerstley borate	4.5	Opacity	Translucent
Whiting	4.4	Color / oxidation	White (32)
Zinc oxide	0.9	/ reduction	Cool white (104)
Yellow ochre	0.9		
	100.0		

G 3064 Oxford Opaque

Oxford feldspar	31	Temperature	C/8–9
Flint	31	Surface @ C/8	
Whiting	16	/ oxidation	Gloss
Georgia kaolin	9	/ reduction	Semigloss
Barium carbonate	5	Fluidity	Little
Dolomite	5	Stain penetration	Darks
Zinc oxide	3	Opacity	Opaque
	100	Color / oxidation	Light olive
Uranium oxide, ground	3		green (159)
		/ reduction	Light tan (39)

G 3065 Jack Pott's Sandstone

Custer feldspar	29.93	Temperature	C/8
Whiting	20.00	Surface @ C/8	Semigloss
Kaolin	9.95	Fluidity	Little
Flint	9.32	Stain penetration	None
Alumina hydrate	6.95	Opacity	Opaque
Zinc oxide	6.64	Color / oxidation	Cream (64)
Zircopax	5.40		
Gerstley borate	4.29		
Rutile	2.27		
Dolomite	1.79		
Yellow ochre	1.59		
Bentonite	0.94		
	99.07		

G 3066 Jack Pott's Sahara

Potash feldspar	24.6	Temperature	C/8
Whiting	22.6	Surface @ C/8	Semigloss
Flint	18.5	Fluidity	None
Kaolin	9.2	Stain penetration	None
Rutile	8.1	Opacity	Opaque
Nepheline syenite	5.8	Color / oxidation	Dark eggshell (60)
Zinc oxide	3.3		
Zircopax	3.3		
Red iron oxide	2.8		
Bentonite	1.8		
	100.0		

Electric-Kiln Firing at Upper-Stoneware Temperatures

This section is intended for potters who fire in the upper-stoneware temperatures in oxidative and neutral—as against reductive—kiln atmospheres. Few potters electric-fire at temperatures above C/12. At such heating levels not only is electric firing slower and more expensive than gas firing, but the electric heating elements will wear out very fast. Most of the 2½-inch-wall kilns that studio potters use, in fact, are not designed for very-high-temperature firing or for *production firing* (one hundred or more firings per year). The tendency of the bricks to decompose because of constant expansion and contraction; the relative frailty of the kiln construction (only painted or stainless steel sheet metal is used); the location of the control box on the wall of the kiln—all contribute to the shortened life expectancy of an electric kiln used too frequently and at very high temperatures. If a studio electric kiln is fired only occasionally (say, a few times a month), however, and if the ceramist becomes familiar with the kiln and its parts and learns both proper maintenance of the kiln and correct techniques for high-temperature firing, then a studio-use kiln should give several years of dependable service.

Both the clay body and glaze contain salts, calcium, sulphur, and other soluble minerals. During bisque- and glaze-firing, these soluble materials are emitted as vapors which travel throughout the kiln and eventually leak out through cracks in the kiln wall and between the door and the kiln wall. They leave minute deposits on kiln shelves, posts, bricks, heating elements, and pots in the kiln. Evidence of this is quite noticeable in a glaze-firing, for the fumes from the glaze on one pot will travel and alight on the pot next to it. Some pots, with or without glazes, will pick up a blush of color from the deposits. Pink spots, for instance, may appear on a high-tin-content glaze, or dark spots may show up on unglazed areas. The glaze activity that is especially harmful to the kiln itself is the bubbling and splattering of some glazes, particularly the soda and boron glazes. The deposits left on the heating elements by the splattering glazes can significantly reduce the life of the elements. Periodic vacuuming of the element's grooves to remove accumulated glaze, as well as clay and dust, will help considerably in minimizing this problem.

During firing, the kiln's elements and electrical components expand and contract. This slight movement causes the electrical connections to corrode, loosen, and arc. Regular cleaning and tightening of the terminals, connections, and electrical plugs will ensure the full amperage necessary for maximum efficiency. Connections in poor condition, on the other hand, will decrease kiln efficiency; kanthal, nichrome, or similar wire will be particularly affected. Each size and length of wire is rated for ohms of resistance, which provides the heat in the kiln. Production, high-temperature, or industrial kilns use silicon carbide elements—rods or bars that serve the same function as the wire elements in the smaller, lower-temperature kilns. Silicon carbide elements are many times more expensive, but they are more durable, more resistant to vapors, maintain better efficiency, and have a longer life.

Because the elements get hotter than do the surrounding bricks and shelves, shelves and pots should never touch the elements or be closer than a half inch from them. Silicon carbide shelves are good conductors of electricity and can cause the elements to short-out if elements and shelves come in contact. A pot placed too close to a heating element will develop a hot spot that can cause the pot to crack or bear a burn mark.

To determine the correct temperature in the kiln so that it can be turned off at a precise time is very important. Overfiring the kiln will result in a too-fluid glaze or in burnt-out color, and underfiring results in an immature surface (sandy) or flat color. Even a ½ cone temperature firing error can have a negative result. The three common and reliable methods of determining the correct turnoff for high-temperature stoneware glazing in the kiln are visual inspection, temperature indicating pyrometer, and kiln guard.

1. *Visual inspection*—Two pyrometric cone sets are placed inside the kiln so that they will be visible through the spy hole. When the firing cone sags, the kiln is turned off.

2. *Temperature indicating pyrometer*—A thermocouple (probe) is inserted through the kiln wall projecting 1 to 2 inches inside the kiln. Wires lead from this probe to the galvanometer, which has a calibrated temperature gauge. When the needle points to the desired temperature, the kiln is turned off either by hand or automatically.

3. *Kiln guard*—A mechanical device designed to shut off the kiln automatically. A high-temperature metal rod extends 1 inch into the kiln from an electrical shut-off control box. At the end is a clamplike device which holds a pyrometric cone. When, in the firing process, the cone reaches the proper temperature, it will soften enough to move the clamp, shutting off the kiln.

Each type of control has a disadvantage. The visual method requires a check of the spy hole every few minutes when the shut-off time is approaching; the pyrometer needs periodic recalibration; and the kiln guard may run into mechanical problems. Firing the kiln several times to adjust the loading, the firing schedule of the glaze, and the glazing technique, as well as keeping careful records of such adjustments, are part of the procedure needed for accurate and consistent firings.

UPPER-STONEWARE-TEMPERATURE GLAZE FORMULAS

The following glaze formulas are for reduction and neutral firing. The characteristics include gloss, semigloss, and matt surfaces, plus transparent, translucent, and opaque appearance. The formulas are for C/8–10, C/11, and C/12.

CONE/9–10

G 3067 Terra Sigillata

Red Horse clay	97	Temperature	C/9
Sodium silicate, dry	3	Surface @ C/9	Slight sheen, semimatt
	100	Fluidity	None
Note: Grind, settle, use top		Stain penetration	None
⅜″ liquid slip		Opacity	Opaque
		Color / oxidation	Chocolate brown
			(74)

G 3068 Clear Gray

Cornwall stone	80	Temperature	C/9–10
Whiting	15	Surface @ C/9	Gloss
Zinc oxide	3	Fluidity	None
Bentonite	2	Stain penetration	All
	100	Opacity	Transparent
		Color / oxidation	Clear grays
		/ reduction	Clear grays

G 3069 Gloss Grayish-White

Cornwall stone	78	Temperature	C/9
Whiting	15	Surface @ C/9	Gloss
Zircopax	7	Fluidity	Little
	100	Stain penetration	All
		Opacity	Transparent
		Color / oxidation	Whitish (48)
		/ reduction	Grayish-white (7)

G 3070 Transparent #98

Kingman feldspar	75.5	Temperature	C/8–10
Zinc oxide	9.0	Surface @ C/9	Gloss
Whiting	8.2	Fluidity	Little
Flint	7.3	Stain penetration	All
	100.0	Opacity	Transparent
		Color / oxidation	Window clear
		/ reduction	Window clear

Note: Apply thin; cracks where thick

G 3071 Porcelain Transparent

Kona A-3 feldspar	75	Temperature	C/9
Dolomite	15	Surface @ C/9	Gloss
Whiting	8	Fluidity	Little
Bentonite	2	Stain penetration	All
	100	Opacity	Transparent
		Color / oxidation	Clear
		/ reduction	Clear

G 3072 Purple

Buckingham feldspar	75.0	Temperature	C/9–10
Whiting	12.5	Surface @ C/9	High gloss
Colemanite	8.7	Fluidity	Little
Bone ash	2.0	Stain penetration	None
Tin oxide	1.0	Opacity	Opaque
Copper carbonate	0.5	Color / oxidation	Purple (107)
Red iron oxide	0.3		
	100.0		

G 3073 Sang-de-Boeuf

Cornwall stone	75.0	Temperature	C/9–10
Whiting	11.0	Surface @ C/9	Gloss
Barium carbonate	10.0	Fluidity	Little
Nepheline syenite	2.5	Stain penetration	All
Tin oxide	1.0	Opacity	Transparent
Red copper oxide	0.5	Color / reduction	Clear, red burst
	100.0		on medium gray

G 3074 Koobation

Cornwall stone	73.5	Temperature	C/9–10
Barium carbonate	12.1	Surface @ C/9	Gloss
Flint	7.2	Fluidity	None
Whiting	4.2	Stain penetration	All
Zinc oxide	3.0	Opacity	Transparent
	100.0	Color / oxidation	Clear, grayish-white
		/ reduction	Clear, grayish-white

G 3075 Broken Tans

Spodumene	70	Temperature	C/9–10
Silica	13	Surface @ C/9	Gloss (broken)
Wollastonite	7	Fluidity	None
Georgia kaolin	5	Stain penetration	All
Zinc oxide	3	Opacity	Transparent
Magnesium carbonate	2	Color / oxidation	Broken tans
	100		(30/36/38)
		/ reduction	Broken tans
			(29/40/38)

G 3076 Sky Flake

Frit #3110 FERRO	67.3	Temperature	C/9
Zinc oxide	24.0	Surface @ C/9	Semimatt
E.P. kaolin	4.8	Fluidity	Little
Green nickel oxide	3.9	Stain penetration	None
	100.0	Opacity	Opaque
		Color / oxidation	Dark olive green (155)

G 3077 Brown-Tan

Albany slip	65	Temperature	C/9–10
Tin oxide	15	Surface @ C/9	Semigloss
Oxford feldspar	10	Fluidity	None
Zinc oxide	10	Stain penetration	Very darks
	100	Opacity	Opaque
		Color / oxidation	Medium-warm brown-tan (38)

G 3078 Keystone Glaze

Keystone feldspar	64	Temperature	C/9
Whiting	20	Surface @ C/9	Semimatt
Georgia kaolin	16	Fluidity	Little
	100	Stain penetration	All
Bentonite	2	Opacity	Transparent
		Color / oxidation	Clear light gray
			On stoneware, (151)
			On porcelain, (8)

Red iron oxide	2 ⎫	Surface @ C/9	
Rutile	6 ⎭	On porcelain	Satin matt
		On stoneware	Dry matt
		Fluidity	Little
		Stain penetration	All
		Opacity	Opaque
		Color / oxidation	On porcelain, mottled tan, orange
			On stoneware, brownish purple (49)

Rutile   5.0 ⎫ Surface @ C/9   Satin matt
Cobalt carbonate   0.9 ⎭
                              Fluidity   Little
                              Stain penetration   None
                              Opacity   Opaque
                              Color / oxidation   Porcelain—light
                                                             blue (87)
                                                  Stoneware—broken
                                               tan and brown (39/74)

G 3079 K-9 Glaze

Kingman feldspar	60.8	Temperature	C/8–9
Dolomite	9.9	Surface @ C/9	Semigloss/matt
Kentucky ball #4	9.1	Fluidity	Some
Barium carbonate	5.5	Stain penetration	None
Georgia kaolin	5.4	Opacity	Opaque
Whiting	4.8	Color / oxidation	Blue marble on
Rutile	4.5		tan (85/39)
	100.0		

G 3080 Semimatt Brown

Cedar Heights Redart	60	Temperature	C/9–10
Frit #311	35	Surface @ C/9	Semimatt
Rutile	5	Fluidity	Some
	100	Stain penetration	None
		Opacity	Opaque
		Color / oxidation	Brown (115)

G 3081 Mouse Gray

Kingman feldspar	59	Temperature	C/9
Red Horse clay	22	Surface @ C/9	Gloss
Whiting	12	Fluidity	None
Flint	5	Stain penetration	All
Zircopax	2	Opacity	Translucent
	100	Color / oxidation	Mouse gray (100)

G 3082 Volcanic

Pumice (volcanic ash)	58	Temperature	C/8–10
Whiting	30	Surface @ C/9	Semimatt/matt
E.P. kaolin	7		smooth
Magnesium carbonate	4	Fluidity	Fluid
Black iron oxide	1	Stain penetration	None
	100	Opacity	Opaque
		Color / oxidation	Broken tans with
			light violets
			(40/112)

G 3083 Transparent Matt

Kona F-4 feldspar	55	Temperature	C/9–10
Zinc oxide	17	Surface @ C/9	Matt
Barium carbonate	14	Fluidity	Some
Whiting	6	Stain penetration	All
Soda ash	4	Opacity	Transparent
Flint	4	Color / oxidation	Clear with slight
	100		frost

G 3084 Custer Porcelain

Custer feldspar	55	Temperature	C/9–10
Tennessee ball #1	13	Surface @ C/9	Gloss
Dolomite	8	Fluidity	None
Whiting	6	Stain penetration	None
Georgia kaolin	6	Opacity	Opaque
Silica	6	Color / oxidation	White (32)
Zircopax	6	/ reduction	White (7)
	100		

G 3085 Eggshell

Custer feldspar	54	Temperature	C/8–10
E.P. kaolin	25	Surface @ C/9	Semigloss/semimatt
Dolomite	18	Fluidity	Little
Whiting	3	Stain penetration	Darks
	100	Opacity	Opaque
		Color / reduction	Off-white (7)

Figure 1

Figure 2

Figure 3

Figure 1 *Vase d'Angers,* porcelain, Sèvres, France. GIFT OF THE FRENCH GOVERNMENT; COURTESY THE FINE ARTS MUSEUM OF SAN FRANCISCO

Figure 2 Maria Longworth Nichols Storer, *Rockwood Tiger's Eye Vase,* 1884, ceramic, 6½″ high. COURTESY THE CINCINNATI ART MUSEUM

Figure 3 John Conrad, *Vase,* mid-temperature clay body, glaze CR 311

Figure 4 John Conrad, *Vase,* earthenware, glaze CR 303, 11″ high

Figure 4

Figure 5

Figure 6   **Figure 7**

Figure 5 Taxile Doat, *Vases,* 1913, crystal glazes. COURTESY THE SMITHSONIAN INSTITUTION

Figure 6 Fulper Pottery, *Oval Vase with Loop Handles,* 1912, green crystal glaze. GIFT OF MR. ROBERT BLASBERG; COURTESY THE SMITHSONIAN INSTITUTION

Figure 7 Taxile Doat, *"Attributes of Thalis" dish,* 19th-20th century, porcelain, 8⅞″ diameter. COURTESY THE ST. LOUIS ART MUSEUM

Figure 8 Taxile Doat, *"Pastoral Poetry" dish,* 19th-20th century, porcelain, 7⅝″ diameter. COURTESY THE ST. LOUIS ART MUSEUM

Figure 9 Taxile Doat, *"The Twins" dish,* Sèvres, 1907, 15¼″ diameter. COURTESY THE ST. LOUIS ART MUSEUM

Figure 10 Matt Daly, *Rockwood Vase,* 1884, crystal glaze. COURTESY ROCKWOOD POTTERY RESTAURANT, CINCINNATI; PHOTO BY RON FORTH

Figure 8

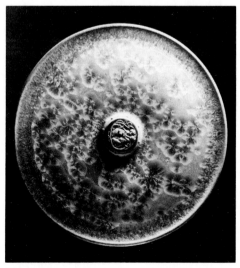

Figure 9

Figure 10

Figure 11

Figure 13

Figure 12

Figure 11 Laura Andreson, *Bowl,* porcelain, pale blue glaze with crystals, 9″ diameter

Figure 12 Laura Andreson, *Vase,* porcelain, tan glaze with cream crystals, 11¾″ high

Figure 13 Rose Cabat, *Bottle,* porcelain, green crystals on "beehive" dark brown crystals, 6″ high

Figure 14   Figure 15

Figure 14   Rose Cabat, *Bottle,* porcelain, blue and brown matt crystalline, 7¾" high

Figure 15   John K. Baker, *Bottle,* porcelain, crystalline glaze, 8" high

Figure 16   John K. Baker, *Bottle,* porcelain, crystalline glaze, 7" high

Figure 16

Figure 17

Figure 18

Figure 19

Figure 20

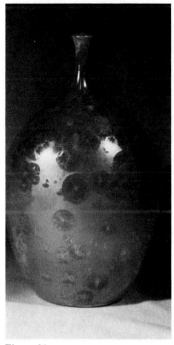 

Figure 21   **Figure 22**

Figure 17   Robert W. Bixler, *Vase,* porcelain, crystal glaze, 8″ high

Figure 18   Jack Feltman, *Bottle,* porcelain, copper on gold crystal glaze, 9½″ high

Figure 19   Jack Feltman, *Bottle,* porcelain, gold on gold crystal glaze, 8″ high

Figure 20   Detail of Figure 19

Figure 21   Jack Feltman, *Bottle,* porcelain, gold crystals on yellow-orange glaze, 7⅜″ high

Figure 22   Jack Feltman, *Bottle,* porcelain, gold crystals on yellow, 11¾″ high

Figure 23   Figure 24   Figure 25

Figure 23 Steven McGovney, *Covered Jar,* porcelain, matt blue crystalline, 11½" high

Figure 24 Steven McGovney, *Covered Jar,* porcelain, off-white crystalline, 13" high

Figure 25 Herbert Sanders, *Oval Vase,* porcelain, yellow crystal, 11½" high

Figure 26 Herbert Sanders, *Lidded Container,* porcelain, green crystal, 10¾" high

Figure 26

Figure 27

Figure 28

Figure 29

Figure 27 Ball of Clay on Masonite Bat

Figure 28 Ball of Clay Centered in a Dome Shape

Figure 29 Cylinder of Clay Pulled

Figure 30 Cylinder of Clay Extended Higher and Straighter

Figure 30

Figure 31

Figure 32

Figure 34

Figure 33

Figure 31 Bottom of Cylinder Being Extended into Bottle Form

Figure 32 Bottom of Bottle Form Being Finished

Figure 33 Top Cylinder Being Necked in by Pressure

Figure 34 Diameter of Cylinder Being Reduced Using a Four-finger Squeeze

Figure 35

Figure 36

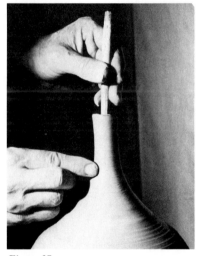

Figure 37

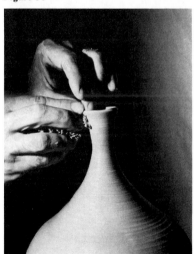

Figure 38

Figure 35 Trimming Off the Irregular Top Edge

Figure 36 One-finger Lift

Fiugre 37 One-finger Lift with Pencil Inside Neck for Support

Figure 38 Flaring Out Lip, and Finishing and Smoothing with Sponge

Figure 39 Rubber Rib Used to Smooth and Shape

Figure 40 Measuring with Calipers

Figure 41 Ball of Clay to Make Catch Basin, and Calipers to Measure the Cylinder

Figure 42 Centered Clay

Figure 39

Figure 40

Figure 41

Figure 42

Figure 43

Figure 44

Figure 43 Cylinder Pulled Up and Brought Out to Correct Measurement

Figure 44 Lip Formed with Remaining Clay

Figure 45 Excess Clay Trimmed from Pot

Figure 45

Figure 46

Figure 47

Figure 48

Figure 49

Figure 46   Catch Basin Remeasured After Drying

Figure 47   Exact Amount to Be Trimmed Determined by New Measurement of Bottle

Figure 48   Top of Catch Basin's Cylinder Sanded Flat After Bisque-firing

Figure 49   Bottle and Basin Surfaces Painted with Mixture of Alumina Hydrate, White Glue, and Water

Figure 50

Figure 51

Figure 52

Figure 53

Figure 50   Pot and Catch Basin Joined

Figure 51   Crystal Glaze Sprayed onto the Pot

Figure 52   Pot and Test Pieces in Test Kiln

Figure 53   Fired Pot and Catch Basin

Figure 54

Figure 55

Figure 56

Figure 57

Figure 54 Pot Freed from Basin

Figure 55 Close-up of Separated Pot and Catch Basin

Figure 56 Finished Crystal-glazed Pot

Figure 57 Detail of Figure 56

Red iron oxide	2.5	Surface @ C/8	Semigloss
		Fluidity	None
		Stain penetration	Darks
		Opacity	Opaque
		Color / oxidation	Very light lavender (79)

Cobalt carbonate	0.3	Surface @ C/8	Semigloss/semimatt
		Fluidity	Little
		Stain penetration	Most
		Opacity	Translucent
		Color / reduction	Grayed-tans with medium browns

Manganese dioxide	1.4	Surface @ C/8	Semigloss/semimatt
		Fluidity	Little
		Stain penetration	None
		Opacity	Opaque
		Color / reduction	Brown with tan speckles

G 3086 MSBA Matt

Kingman feldspar	54.0	Temperature	C/9
Whiting	15.4	Surface @ C/9	Semimatt
Georgia kaolin	14.7	Fluidity	None
Barium carbonate	12.2	Stain penetration	All
Flint	3.7	Opacity	Transparent
	100.0	Color / oxidation	Clear gray
		/ reduction	Clear gray

Red iron oxide	6.0	Surface @ C/9	Semimatt
		Fluidity	None
		Stain penetration	None
		Opacity	Opaque
		Color / reduction	Army green on rust (137/75)

Cobalt carbonate	0.3	Surface @ C/9	Dry matt
Nickel carbonate	2.0	Fluidity	None
		Stain penetration	Most
		Opacity	Translucent
		Color / reduction	Gray and tans
			(104/27)

G 3087 Kona A-3

Kona A-3 feldspar	49	Temperature	C/9–10
Georgia kaolin	23	Surface @ C/9	Matt
Whiting	20	Fluidity	Little
Talc	4	Stain penetration	None
Bone ash	2	Opacity	Opaque
Tin oxide	1	Color / reduction	Broken mottled
Copper carbonate	1		purple/mottled green
	100		(9/137)

G 3088 Clear Broken Gray/Brown

Lepidolite	49	Temperature	C/9–10
Barium carbonate	19	Surface @ C/9	Semigloss
Flint	15	Fluidity	None
Georgia kaolin	9	Stain penetration	All
Whiting	6	Opacity	Transparent
Zircopax	2	Color / oxidation	Clear, broken light
	100		browns
		/ reduction	Clear, broken grays
			and browns

G 3089 Semimatt

Oxford feldspar	47.6	Temperature	C/9–10
E.P. kaolin	25.3	Surface @ C/9	Semimatt
Whiting	21.8	Fluidity	None
Talc	3.5	Stain penetration	None
Bone ash	1.8	Opacity	Opaque
	100.0	Color / reduction	Blue-gray white
			(104)
Red iron oxide	6.0	Color / reduction	Rust (77)

G 3090 Light Violet-Blue

Kingman feldspar	47.5	Temperature	C/9–10
Georgia kaolin	24.8	Surface @ C/9	Semimatt
Dolomite	22.3	Fluidity	Little
Whiting	3.5	Stain penetration	None
Calcium chloride	1.5	Opacity	Opaque
Black cobalt oxide	0.5	Color / reduction	Light violet-blue
Chrome oxide	0.1		(112/88)
	100.2		

G 3091 New Shaner Base

Kingman feldspar	47.5	Temperature	C/9–10
E.P. kaolin	20.2	Surface @ C/9	Smooth matt
Whiting	19.2	Fluidity	Little
Bone ash	9.1	Stain penetration	All
Talc	4.0	Opacity	Translucent
	100.0	Color / reduction	On stoneware, light gray (104)
			On porcelain, pink tinge

Red iron oxide	4.0 ⎫	Surface @ C/9	Satin matt
Rutile	1.0 ⎭	Fluidity	Little
		Stain penetration	All
		Opacity	Translucent
		Color / reduction	On stoneware, dark brown (74)
			On porcelain, greenish-gray (159)

Cobalt carbonate	0.5 ⎫	Surface @ C/9	Satin matt
Copper carbonate	1.5 ⎬	Fluidity	None
Rutile	1.0 ⎭	Stain penetration	Most
		Opacity	Opaque
		Color / reduction	On stoneware, gray-blue (102)
			On porcelain, purple (120)

G 3092 MacIntosh

Kingman feldspar	46.8	Temperature	C/8–9
Talc	13.6	Surface @ C/8	Gloss
Flint	15.3	Fluidity	Some
Gerstley borate	11.7	Stain penetration	All
Dolomite	8.1	Opacity	Transparent
E.P. kaolin	4.5	Color / reduction	White (pale green-gray)
	100.0		
		/ oxidation	Clear, very pale greenish-gray

G 3093 Broken White/Tan L.A. D-2

Kingman feldspar	45	Temperature	C/9–10
Georgia kaolin	35	Surface @ C/9	Matt
Dolomite	20	Fluidity	None
	100	Stain penetration	Most
		Opacity	Translucent
		Color / oxidation	Broken white (7) and tan
		/ reduction	Broken white (7) and tan

G 3094 Multicolor Matt

Kingman feldspar	44.0	Temperature	C/9
E.P. kaolin	22.5	Surface @ C/9	Satin matt
Dolomite	19.9	Fluidity	Little
Whiting	13.6	Stain penetration	All
	100.0	Opacity	Opaque
		Color / reduction	On stoneware, mottled light yellow, gray, white
			On porcelain, light gray/white specks

G 3095 Otto's Hare's Fur

Cornwall stone	43.8	Temperature	C/9–10
Colemanite	43.8	Surface @ C/9	Broken gloss/matt
Rutile	8.8	Fluidity	Little
Red iron oxide	1.8	Stain penetration	Most
Alumina hydrate	1.8	Opacity	Transparent
	100.0	Color / reduction	Clear light green

G 3096 Semigloss White

Oxford feldspar	43.1	Temperature	C/9–10
Flint	22.9	Surface @ C/9	Semigloss
Talc	12.5	Fluidity	Little
Colemanite	9.8	Stain penetration	Darks
Dolomite	7.5	Opacity	Opaque
Kentucky ball	4.2	Color / reduction	White (32)
	100.0		

G 3097 Glossy (Laura Andreson)

Kingman feldspar	42.4	Temperature	C/9–10
Flint	26.5	Surface @ C/9	Gloss
Colemanite	8.8	Fluidity	Little
Dolomite	8.8	Stain penetration	All
Barium carbonate	4.4	Opacity	Transparent
Tin oxide	2.7	Color / oxidation	Clear light gray
Whiting	2.7	/ reduction	Flame
Zinc oxide	1.8		
E.P. kaolin	1.5		
Copper carbonate	0.4		
	100.0		

G 3098 Kingman Feldspar

Kingman feldspar	41.7	Temperature	C/9
Whiting	20.0	Surface @ C/9	Gloss
Flint	10.8	Fluidity	Little
Georgia kaolin	10.8	Stain penetration	Darks
Rutile	7.5	Opacity	Opaque
Colemanite	5.0	Color / reduction	Broken tans and
Black iron oxide	4.2		light greenish-
	100.0		ochre (39/17)

G 3099 Alfred University Yellow

Cornwall stone	40	Temperature	C/8–10
Georgia kaolin	25	Surface @ C/9	Matt
Dolomite	15	Fluidity	Some
Whiting	10	Stain penetration	Most
Flint	10	Opacity	Opaque
	100	Color / reduction	Light gray-blue
			(100)

Red iron oxide	3.0		
Rutile	5.0	Surface @ C/9	Matt
Bentonite	2.0	Fluidity	Thick fluid
		Stain penetration	None
		Opacity	Opaque
		Color / reduction	Marbled tans (39)
			and yellows

G 3100 Warm White Mother-of-Pearl

Flint	40	Temperature	C/8–10
Zinc oxide	20	Surface @ C/9	High gloss
Whiting	20	Fluidity	Fluid
Sodium carbonate	14	Stain penetration	The very darks
Titanium dioxide	3	Opacity	Opaque
Kaolin	3	Color / oxidation	Warm white
	100		mother-of-pearl

G 3101 Custer White

Custer feldspar	40	Temperature	C/9–10
Flint	20	Surface @ C/9	Satin matt
Talc	15	Fluidity	Little
Colemanite	11	Stain penetration	Most
Dolomite	7	Opacity	Opaque
Tennessee ball #1	7	Color / reduction	White (7)
	100		
Bentonite	2		

G 3102 Oxford Blue-Gray

Oxford spar	40	Temperature	C/9
Georgia kaolin	20	Surface @ C/9	Gloss
Whiting	10	Fluidity	Little
Dolomite	10	Stain penetration	All
Talc	10	Opacity	Transparent
Flint	10	Color / reduction	Clear, slight blue-
	100		gray

G 3103 Light Gray (Peter Voulkos)

Nepheline syenite	39.8	Temperature	C/9
E.P. kaolin	20.8	Surface @ C/9	Smooth matt
Dolomite	14.0	Fluidity	Little
Flint	11.7	Stain penetration	All
Whiting	7.7	Opacity	Opaque
Zircopax	4.0	Color / reduction	Light gray (104)
Zinc oxide	2.0		
	100.0		

G 3104 Keystone White

Keystone feldspar	39	Temperature	C/9–10
Georgia kaolin	27	Surface @ C/9	Matt
Whiting	16	Fluidity	None
Talc	15	Stain penetration	Darks
Flint	3	Opacity	Opaque
	100	Color / reduction	White (7)

G 3105 Kona Multicolor

Kona F-4 feldspar	38.9	Temperature	C/9
Dolomite	21.6	Surface @ C/9	Satin matt
Kentucky ball #4	21.6	Fluidity	Little
Nepheline syenite	8.5	Stain penetration	Most
Whiting	4.3	Opacity	Opaque
Tin oxide	4.0	Color / reduction	Mottled ochre,
Bentonite	1.1		orange, browns
	100.0		
Cobalt carbonate	0.7	Color / reduction	Broken purple,
			lavender, tan,
			dark brown, blues

Black iron oxide	1.0 ⎫	Color / reduction	Broken light
Manganese dioxide	1.0 ⎭		tan, browns (39/34)

G 3106 Peach Bloom (Laura Andreson)

Kingman feldspar	38.0	Temperature	C/9–10
Silica	33.9	Surface @ C/9	Gloss
Whiting	14.3	Fluidity	Little
E.P. kaolin	6.5	Stain penetration	All
Barium carbonate	5.1	Opacity	Transparent
Zinc oxide	2.2	Color / reduction	Clear light gray
	100.0		(cracks)

Tin oxide	1.0 ⎫		
Copper carbonate	1.0 ⎬	Surface @ C/9	High gloss
Red iron	1.0 ⎭	Fluidity	Little
		Stain penetration	All
		Opacity	Transparent
		Color / reduction	Light broken purple/
			light peach bloom

G 3107 Century Gray

Nepheline syenite	37	Temperature	C/9
Flint	34	Surface @ C/9	Gloss
Dolomite	11	Fluidity	None
Whiting	9	Stain penetration	All
Barium carbonate	7	Opacity	Transparent
E.P. kaolin	2	Color / reduction	Clear, gray
	100		

G 3108 Persimmon

Flint	36.3	Temperature	C/9
Kingman feldspar	29.9	Surface @ C/9	On porcelain,
Whiting	15.0		high gloss
Red iron oxide	11.3		On stoneware,
Georgia kaolin	7.5		satin matt
	100.0	Fluidity	Little
		Stain penetration	None
		Opacity	Opaque
		Color / reduction	On porcelain, dark
			brownish-black (73)

On stoneware, brownish-persimmon

G 3109 Pink Tinge

Custer feldspar	36	Temperature	C/9
Whiting	25	Surface @ C/9	Satin matt
Georgia kaolin	20	Fluidity	Little
Spodumene	14	Stain penetration	Darks
Gerstley borate	5	Opacity	Opaque
	100	Color / reduction	Light gray with
			pink tinge (151)

G 3110 Custer Gray

Custer feldspar	35	Temperature	C/9–10
Kentucky ball #4	33	Surface @ C/9	Gloss
Whiting	19	Fluidity	None
Flint	13	Stain penetration	All
	100	Opacity	Transparent
		Color / reduction	Clear gray

G 3111 Tenmoku

Wollastonite	33.6	Temperature	C/9
Silica	21.3	Surface @ C/9	Gloss
Lincoln clay	18.2	Fluidity	None
Red iron oxide	15.4	Stain penetration	Darks
Keystone feldspar	7.7	Opacity	Opaque
Zinc oxide	3.8	Color / reduction	Brownish-
	100.0		purple (107)

G 3112 Pauline

Flint	33.3	Temperature	C/9
Potash feldspar	22.9	Surface @ C/9	Gloss with
Dolomite	14.6		matt areas
E.P. kaolin	12.4	Fluidity	None
Whiting	11.1	Stain penetration	All
Colemanite	4.2	Opacity	Transparent
Barium carbonate	0.9	Color / reduction	Clear
Tin oxide	0.6		
	100.0		
Nickel	7.0	Surface @ C/9	Matt
		Fluidity	None
		Stain penetration	Darks
		Opacity	Opaque
		Color / reduction	Medium blue,
			grayish-green

G 3113 Wildenhain White

Kingman feldspar	33.3	Temperature	C/9–10
Dolomite	22.2	Surface @ C/9	Semigloss
E.P. kaolin	22.2	Fluidity	None
Flint	22.2	Stain penetration	All
	99.9	Opacity	Translucent
		Color / reduction	White clear (7)

G 3114 L.A. 52

Kingman feldspar	30	Temperature	C/9–10
Ball clay	30	Surface @ C/9	Matt
Whiting	20	Fluidity	Little
China kaolin	10	Stain penetration	Most
Silica	10	Opacity	Translucent
	100	Color / reduction	Broken tans (28/40)

G 3115 Cool White Buttermilk

Custer feldspar	27	Temperature	C/9–10
Flint	22	Surface @ C/9	Semigloss/
Talc	14		semimatt
Colemanite	9	Fluidity	None
Whiting	8	Stain penetration	Most
Zircopax	7	Opacity	Translucent
Georgia kaolin	7	Color / reduction	Cool white
Dolomite	6		buttermilk (7)
	100		

CONE/11

G 3116 Whitish Porcelain

Flint	44.3	Temperature	C/11
China clay, calcined	24.2	Surface @ C/11	Gloss
Kingman feldspar	16.2	Fluidity	Little
Whiting	11.6	Stain penetration	All
Georgia kaolin	3.7	Opacity	Transparent
	100.0	Color / reduction	Whitish (8)

G 3117 Kingman Potpourri

Kingman feldspar	43.9	Temperature	C/11–13
Whiting	21.1	Surface @ C/11	Gloss
Flint	11.4	Fluidity	Some (thick)
Georgia kaolin	11.4	Stain penetration	None
Rutile	7.9	Opacity	Opaque
Red iron oxide	4.3	Color / reduction	Broken blues, tans,
	100.0		light browns
			(101/35/28)

G 3118 Gray Celadon

Flint	32.8	Temperature	C/11
Feldspar potash	27.1	Surface @ C/11	Gloss
Whiting	19.6	Fluidity	Little
Georgia kaolin	19.6	Stain penetration	All
Red iron oxide	0.9	Opacity	Transparent
	100.0	Color / oxidation	Slight gray celadon

CONE/12

G 3119 Clear Porcelain

Flint	48	Temperature	C/12
Whiting	26	Surface @ C/12	Gloss
Georgia kaolin	26	Fluidity	None
	100	Stain penetration	All
		Opacity	Transparent
		Color / oxidation	Clear

G 3120 Slight White L.B.S.I.I.P.

Flint	33.4	Temperature	C/12
Buckingham feldspar	26.4	Surface @ C/12	Gloss
Georgia kaolin	20.2	Fluidity	Some
Whiting	17.4	Stain penetration	All
Magnesium Carbonate	1.5	Opacity	Transparent
Zinc oxide	1.1	Color / oxidation	Slight white (7/8)
	100.0		(slight cracking)

G 3121 Clear Porcelain

Silica	31	Temperature	C/12–14
Kingman feldspar	26	Surface @ C/12	Gloss
Whiting	23	Fluidity	Little
China kaolin	20	Stain penetration	All
	100	Opacity	Transparent
		Color / reduction	Clear

VII

Aventurine, Crystal, and Crystalline Matt Glazes

Aventurine, crystal, and crystalline matt glazes have excess minerals that form crystals in the glaze matrix. The first known crystal glaze was found in China; it was called Fat-shan Chun, after the locality in which it was made. Some examples of Chun glaze date back to the late Ming Dynasty (1368–1644). Many of the Chun glazes have attractive soft blue surfaces with a silken luster. There are even a few rare pieces in which the glaze has flowed to the center of the dish, creating a deep pool of glass with a crystal mass of vivid color and pattern. A slightly different type of crystal glaze—in which fewer individual crystals are visible—had appeared, during the Tang Dynasty (618–907), in the Fukien province of southern China. This glaze, known as *oil spot,* is dark, with silvery iridescent spots produced by iron and calcium.

In Europe, some of the earliest scientific studies of crystal glazes were carried out in France. One of the first studies involving zinc crystals was done by Ebelman (1847–1852). Charles Lauth and G. Dutailly, using high-zinc, low-alumina glazes, conducted the first extensive research of crystal glazes at the Manufacture Nationale de Sèvres (Figure 1). Lauth's most successful crystal formula was as follows:

Pegmatite (feldspar)	54.7
Zinc oxide	17.7
Sand (silica)	15.7
Lime (whiting)	11.8
	99.9

Between 1880 and 1935 a number of comprehensive studies in aventurine, crystal, and crystalline matt glazes were undertaken in France, Denmark, Germany, Britain, and the United States. Many of these studies are listed in the bibliography.

Types of Crystal Glazes

Crystal and aventurine glazes contain crystals that are visible to the naked (unaided) eye, but in crystalline matt the crystals are too small to see, even with a 40-power scope. However, the crystals give the glaze the matt surface and can be felt. In the 1950s, Cullen Parmelee classified crystals into two primary types:

1. *Macrocrystalline* in which the crystals are large enough to be readily distinguished by the naked eye. There are two kinds:
    a. Aventurine glazes, where the crystals are sufficiently large and separated as to be easily observed as minute scales or "flitters" suspended below the surface of the glaze; or they may be very small and so numerous as to reflect light rays from the internal surfaces of the suspended crystal, producing an effect similar to the mineral "cat's eye."
    b. Crystalline glazes (the typical crystal glaze), where the surface may be covered completely or partially with well-developed individual crystals (some, perhaps, projecting above); or where the crystals

may be below the surface, immersed in the glass matrix. The crystals are usually in clusters, frequently covering considerable area.

2. *Microcrystalline* in which the crystals are not individually visible except under magnification. The typical example is the matt glaze. This has a surface which, in the best examples, has the soft, pleasing appearance of a glimmering, resinous luster similar to that exhibited by jade and chert. Less satisfactory surfaces are semi-gloss, dull, and greasy.*

This author has chosen to use a simpler classification method, though it is technically the same as Parmelee's, to serve as a guide to make identification easier:

Aventurine: A glaze with minute crystals in suspension that are large enough to be readily observed, and that sparkle and reflect light.

Crystal: A glaze in which true crystal is on the surface, separated and easily distinguishable as individual crystals or clusters of crystals.

Crystalline matt: A glaze whose entire surface is covered with minute crystals that are not individually visible except under magnification.

General Techniques for Working with the Three Glaze Types

Aventurine, crystal, and crystalline matt glazes have been produced in a wide variety of compositions and temperatures; and a review of the published research shows that all the glazes developed have a common feature: All have minerals in their composition that separate out from the molten matrix. The minerals are added to a basic, slightly fluid glaze; they appear throughout the whole mass (a colloid). During the soaking period they collocate around a nucleus, forming a crystal.

Nearly every mineral used in ceramics can be found in nature in its crystal form. In crystal glazes, however, the mineral is added up to and beyond the supersaturation point; the excess develops into the crystals. Iron, chrome, uranium, and manganese oxide will

* Cullen W. Parmelee, *Ceramic Glazes* (Chicago: Industrial Publications, 1951), p. 189.

produce aventurine crystals. Zinc, titanium, and manganese oxide will form crystals in both crystalline matt and crystal glazes.

The following conditions and procedures apply to the preparation, application, and firing of crystal-type glazes:

1. The amount of alumina is under 10 percent.
2. The glaze fluidity ranges from very fluid to a medium flow.
3. The majority of the glazes are of the alkaline type.
4. The glaze coating is thicker than average.
5. Spraying the glaze produces the best results.
6. Fritted (part or all) glazes produce crystals more easily than do their counterpart raw glazes.
7. A soaking period is necessary to heal pits and to allow the crystals to develop.
8. Fresh glaze produces more crystals than glazes stored a long time. Prepare just enough glaze to be used at one time.
9. The glazes must be well mixed, screened, and sometimes ball-milled.
10. The early candling stage of the glaze fire must be slow enough to permit the removal of physical and chemical water.
11. Higher-than-normal bisque-firing of the ware is necessary when soluble glazes are used; otherwise, the soluble fluxes enter the clay body and fuse, forming a hard crust that keeps the clay body from shrinking in the firing. This can cause the pot to break.
12. Aventurine and crystal glazes need a neutral or oxidative firing and soaking; otherwise, the glaze is under tension, and, consequently, the growth of the crystals may be inhibited.

Aventurine Glazes

The name *aventurine* is given to glazes—similar in appearance to natural minerals of the quartz and feldspar family—which contain mica or hematite flitters that sparkle when held in a bright light. This special type of glaze, also known as *flitter, spangle, cat's-eye, goldstone,* and *tiger's-eye,* contains individual crystals isolated from one another in the matrix, giving rise to the characteristic appearance of flitters suspended in the glaze. To achieve an aventurine effect, metallic oxides are added to a glaze in such quantities that, when heated to the molten state, the glass matrix becomes saturated. As the glaze cools, a condition of supersaturation develops, with the ultimate precipitation of the excess metallic oxides into thin, plate-like crystals (flitters) suspended in the glaze.

Aventurine glazes are not new; they were created several centuries ago in the Orient. T. Wohler produced such glazes in Germany around 1845, and a few years later, Ebelman did his studies on them. During the 1880s, the major ceramic factories in Copenhagen, Sèvres, Berlin, and Cincinnati—as well as smaller factories and individual potters—became interested in the shape of the pot as pure form and concentrated on the glaze treatment. They sought to imitate the glazes of the Orient, such as celadon, sang-de-boeuf, tenmoku (hare's fur, oil spot), flambé, as well as aventurine and crystal. After the chemists in the Copenhagen and Berlin factories had developed the crystal glazes, commercial production was started. In 1880, Maria Longworth Storer founded the Rookwood pottery in Cincinnati. Further extensive study was done, and in 1884 the first American ceramics with aventurine glaze were introduced (Figure 2). Rookwood forms and glazes achieved an international reputation. In 1893, at the Columbian Exposition, the Rookwood pottery exhibited crystal-glazed ware that elicited the interest of ceramists worldwide. Rookwood's greatest triumph, though, was at the Paris Exhibition, where the pottery received a gold medal in 1889.

Fifty years later Mackler tried Wohler's formula, but could not obtain satisfactory results. He experimented with various approaches, and finally developed a similar glaze, one in which the calcium that was the principal flux produced a dark-colored glaze with flitters. Of the many formulas he tested, his best is as follows:

C/09 Mackler's Aventurine

0.25 K_2O
0.25 Na_2O 0.75 B_2O_3 2.25 SiO_2
0.50 CaO

Flint	37.29
Borax	26.38
Potassium nitrate	13.95
Whiting	13.81
Boric acid	8.53
	99.96
Iron oxide	20.00

THE GLAZE

Aventurine is a transparent gloss glaze which achieves its effect by the formation of small, bright, and sometimes colored crystals that are suspended in the glaze. These crystals—generally hematite (Fe_2O_3)—reflect light in a sparkling way and show best under a bright light. The most successful aventurines are produced with lead-, boron/lead-, or soda-based glazes. In general, most glazes will absorb from 3 to 10 percent of a metallic oxide such as iron, chrome, uranium, nickel, copper, or cobalt. A greater amount will crystallize under slow cooling and, as in the case of aventurine glazes, form visible crystals. The most common oxide used is iron, alone or with other metals. Some compositions may use copper, chrome, and uranium. If too much oxide is present, the excess will form on the glaze surface as coarse crystals or as a metallic film. Most of the formulas included in this book are for the earthenware and mid-range temperatures. Various statements in the literature suggest that stoneware temperatures can possibly be used as well.

Considerations for developing aventurine glazes are as follows:
1. Decreasing the amount of lead and increasing the soda in the formula will increase the number of flitters, but the high gloss will decrease.
2. The best fluxing agents are soda, potash, boron, and lead, with calcium the least effective.

3. Increasing the iron, chrome, uranium, manganese oxide, or other metallic oxide will produce larger and more numerous flitters, but too much will produce a stony finish or bubbles.
4. Very little alumina is used in the glaze.
5. Thin coatings will be sandy. The best results are obtained with medium to heavy glaze coating.
6. Many of the formulas will produce a slight craze on most clay bodies.

Glaze Application

Many of the components of aventurine glazes, such as soda, potassium, and boron, are water-soluble. Only as much glaze as can be used immediately is mixed at one time. Some glazes settle very quickly and in less than one hour can become caked. They will then need to be broken up and reground before they can be used. This, in fact, is one of the major reasons why frits are used in aventurine and other crystal glazes.

The glaze materials are weighed and enough water is added to reach a pouring consistency. The mixture is then ground and screened through a 60-mesh screen. Regardless of the application method used, it is necessary to grind all glaze materials. Small amounts can be hand-ground in a mortar and pestle, large amounts in a ball mill. Unfortunately, ball-mill grinding requires more water than is needed for glaze-application consistency.

Spraying is the preferred application method, with pouring, dipping, and painting being second-best alternatives. A small amount of glaze binder (CMC or gum arabic) is added to the glaze mix to compensate for the extra thickness of the glaze and its tendency to chalk or to flake off the pot. Some aventurine glazes are superfluid, while others flow only enough to form flitters. In general, the amount of glaze applied is about the thickness of a dime, with a little more at the top and edges, for there is a tendency for the glaze to become very thin. For glazes that are very fluid, a catch basin is recommended

(Figure 50). With practice, the proper amount of glaze can be applied so that the glaze will run slightly and still form crystals.

FIRING

The firing temperature depends upon the glaze composition, the shape of the pot, the clay body, and the tendencies of the crystal formers in the glaze. The temperature should be such that the glaze flows slightly, matures completely, and takes on a smooth surface. When the firing temperature has not been high enough or has been turned off too suddenly, the glaze will frequently be too thick, taking on an irregular, immature, or sandy surface. If it is overfired, the glaze will be runny, leaving thin spots, especially at top edges. Some glazes will form flitters very easily and will not need a very long soaking period. Glazes with a weak or slow tendency to form crystals should be given a longer time to soak.* (Soaking or slow-cooling is essential for the development of large flitters.) The glaze is soaked at the peak temperature for twenty to fifty minutes; or, depending on the glaze, the temperature is lowered about 100 degrees and held for one hour. If soaked too long, the glaze will form a crystalline matt surface. When experimenting with aventurine glazes, in any event, one should not come to a hasty conclusion, based on the results of only one firing; rather, the test pieces should be fired at different temperatures. Moreover, several pieces must be used for firing each test glaze at the various temperatures. Test glazes should also be *refired,* for some of the most effective crystal glazes are the result of a second firing. For some unknown reason, refiring will sometimes produce flitters not generated in the first firing.

A range of colors, from red to brown to black, may be obtained by varying the kiln atmosphere after the soaking period. If a pot, especially one with a heavy lead-based glaze, is not protected from the kiln's gases, the glaze will become dull black with a metallic luster caused by reduction of the ferric oxide. Applying a thin coat of lead-based, high-gloss glaze to reduced glazes and refiring in an oxidative atmosphere will produce a maroon color.

---

* *Soaking* is the process of holding the temperature of the kiln steady for a period of time to permit the glaze to mature, bubbles to break and heal, and the crystals to form.

Clay Body

Since most aventurine glazes are dark, the color of the clay body is not critical. Any clay body that vitrifies at the glaze-maturing temperature can be used. Because the majority of aventurine glazes are in the earthenware temperature range, the clay body would ordinarily be in this range. For mid-temperature and stoneware aventurine glazes, the clay body is bisque-fired at 1900°F.

AVENTURINE GLAZE FORMULAS

Each potter mixes and applies the glaze, and fires the kiln, slightly differently from the way other potters do. One potter may prefer to dip twice in a thin glaze; another will single-dip in a thicker mix; a third may spray, while yet a fourth will paint the glaze onto the body. Such individual differences will also be apparent in kiln firing, for there are intricate variations in electric and gas firings: the potter may fire to light, middle, or flat cone, or may choose one or another type of body reduction. What is more, some kilns have as much as a two-cone difference between top and bottom temperatures. These variations in personal preference and in facilities available are often reflected in the characteristics of the glazes produced. For example, if two potters selected the same glaze formula and then each mixed it, applied it to one of his or her own pots, and fired the pot, a side-by-side comparison of the finished pots might show results so dissimilar that one might not believe that the two had started with the same glaze formula.*

Figures 3 (glaze CR 311) and 4 (glaze CR 303) are photographs of aventurine glazes.

The formulas given are for C/010 through C/9; semimatt, semigloss, and gloss finishes; bases of fritted and of raw materials; and semiflowing to fluid consistency. Please note that some glazes contain unfritted and/or excessive amounts of lead and could be toxic when used on pottery destined for food serving or storage. Such glazes are marked here. Also, see the section Toxicology and Safety in Ceramics.

* It is for these reasons that it is essential to test each glaze before using it on a good pot.

CR 300 Broken Red and Black TOXIC

Red lead	73.5	Temperature	C/010–06
China clay	9.8	Surface @ C/06	Semigloss
Flint	6.9	Fluidity	Some
Soda ash	4.9	Stain penetration	None
Whiting	3.9	Opacity	Opaque
Zinc oxide	1.0	Color / oxidation	Broken reds and
	100.0		blacks
		Crystal	Gold flitters

CR 301 Modeled Colors TOXIC

White lead	65	Temperature	C/010–06
Silica	15	Surface @ C/06	Gloss
Oxford feldspar	13	Fluidity	Some
Rutile	7	Stain penetration	None
	100	Opacity	Opaque
		Color / oxidation	Green-brown (138)

Cobalt carbonate	6	Surface @ C/06	Matt
		Fluidity	None
		Opacity	Opaque
		Color / oxidation	Black (113)
		Crystal	Some minute silver flecks

Red iron oxide	20	Surface @ C/06	Matt
		Fluidity	Fluid
		Opacity	Opaque
		Color / oxidation	Dark chocolate (115)
		Crystal	Very minute silver flecks; glitters like sugar. Apply evenly; best if fritted.

Nickel oxide	12	Surface @ C/06	Semimatt
		Fluidity	Little
		Opacity	Opaque
		Color / oxidation	Broken olive browns and greens
		Crystal	Some minute silver flecks

CR 302 Silver and Gold Flitters

Borax	61	Temperature	C/010
Flint	30	Surface @ C/010	Matt
Black iron oxide	9	Fluidity	None
	100	Stain penetration	None
		Opacity	Opaque
		Color / oxidation	Dark chocolate (115)
		Crystal	Many silver and gold flitters

CR 303 Mirror Black
(See Figure 4)

Frit #25 HOMMEL	75.2	Temperature	C/07
Red iron oxide	16.5	Surface @ C/07	Smooth gloss
Ball clay	5.5	Fluidity	Fluid
Cobalt carbonate	2.8	Stain penetration	None
	100.0	Opacity	Opaque
		Color / oxidation	Mirror true black (113)
		Crystal	Some minute silver flecks; scattering of pinhead silver surface crystal; some cracks

CR 304 Gloss/Matt

Frit #5301	70.1	Temperature	C/07
Red iron oxide	16.8	Surface @ C/07	Gloss and matt
Pearl ash	5.6	Fluidity	Some
Copper oxide	3.8	Stain penetration	None
Ball clay	3.7	Opacity	Opaque
	100.0	Color / oxidation	True black and metallic black
		Crystal	Gloss—minute silver flecks Matt—pinhead surface crystals

CR 305 True Black

Lead bisilicate	61.9	Temperature	C/07–06
Copper carbonate	15.5	Surface @ C/07	Gloss
Pearl ash	10.3	Fluidity	Some
Bismuth subnitrate	7.2	Stain penetration	None
Boric acid	5.2	Opacity	Opaque
	100.1	Color / oxidation	True black
		Crystal	Some minute silver flitters; some cracks

CR 306 Nickel Aventurine

Calcined borax	59	Temperature	C/08
Flint	22	Surface @ C/08	Gloss
Boric acid	11	Fluidity	Little
White lead	5	Stain penetration	None
Ball clay	3	Opacity	Opaque
	100	Color / oxidation	Spring green (127)
Nickel	11	Crystal	Scattered green flitters

CR 307 Rookwood Goldstone TOXIC

White lead	58.0	Temperature	C/04
Soda feldspar	24.3	Surface @ C/04	High gloss
Flint	12.1	Fluidity	Some
Red iron oxide	3.3	Stain penetration	None
Whiting	2.3	Opacity	Opaque
	100.0	Color / oxidation	Dark brown, red-brown where thick (74)
		Crystal	Gold flecks

Note: Apply a thin wash of potassium dichrome and water over the pot, then the glaze over the wash.

CR 308 Black TOXIC

Red lead	58	Temperature	C/04
Flint	20	Surface @ C/04	Gloss
Manganese carbonate	9	Fluidity	Fluid
Cornwall stone	9	Stain penetration	None
Red iron oxide	4	Opacity	Opaque
	100	Color / oxidation	Mirror black (113)
		Crystal	Some minute silver flecks

CR 309 Chrome Aventurine

Colemanite	53	Temperature	C/04
Plastic vitrox	25	Surface @ C/04	Gloss
Nepheline syenite	15	Fluidity	None
Chrome oxide	7	Stain penetration	None
	100	Opacity	Opaque
		Color / oxidation	Chrome green (124)
		Crystal	Some medium blue flitters (99)

CR 310 Silver Flitters TOXIC

Red lead	39	Temperature	C/02–4
Borax	20	Surface @ C/02	Soft matt; where thick, gloss
Red iron oxide	20		
Silica	15	Fluidity	Fluid where thick
Kaolin	6	Stain penetration	None
	100	Opacity	Opaque where thick
		Color / oxidation	Medium brown
		Crystal	Crystalline matt with small silver flitters; some cracks. Apply medium thick.

CR 311 Green/Black
(See Figure 3)

Silica	44	Temperature	C/2–6
Lithium	30	Surface @ C/5	Broken smooth matt;
Chrome oxide	13		some gloss
Potassium nitrate	8	Fluidity	Fluid
Barium carbonate	5	Stain penetration	None
	100	Opacity	Opaque
		Color / oxidation	Gloss—broken
			dark green
			Matt—broken black/green
		Crystal	Gloss—silver flitters
			Matt—sugar surface flitters

CR 312 Dark Brown

Frit #3304 FERRO	80.5	Temperature	C/3–5
Black iron oxide	15.0	Surface @ C/3	Gloss
Calcined kaolin	4.5	Fluidity	None
	100.0	Stain penetration	None
		Opacity	Opaque
		Color / oxidation	Dark brown (154)
		Crystal	Silver flitters

CR 313 Fritted Aventurine
(Frit formula used in the following two glazes)

Flint	35.4
White lead	26.1
Borax	16.1
Soda nitrate	15.8
Red iron oxide	6.5
	99.9

Glaze 1

Frit	49.9	Results of the two glazes are the same
Red iron oxide	23.6	Temperature	C/4
White lead	16.6	Surface @ C/4	Gloss
Flint	7.1	Fluidity	Some flow
Tennessee ball	2.8	Stain penetration	Darks
	100.00	Opacity	Translucent

Glaze 2

Color / oxidation — Brown-black
Crystal — Large flitters

Frit	40.6
Flint	24.5
Red iron oxide	19.2
White lead	13.5
Tennessee ball	2.3
	100.1

CR 314 Barium Aventurine

Oxford feldspar	50	Temperature	C/5
Barium carbonate	20	Surface @ C/5	Matt
China clay	10	Fluidity	Little
Frit #3191	10	Stain penetration	None
Cobalt oxide	6	Opacity	Opaque
Manganese carbonate	4	Color / oxidation	Blue-black
	100	Crystal	Minute silver specks

CR 315 KC-NST

Potash feldspar	54.3	Temperature	C/9
Flint	22.4	Surface @ C/9	Gloss
Red iron oxide	12.8	Fluidity	Some
Whiting	6.4	Stain penetration	None
Kaolin	4.2	Opacity	None
	100.1	Color / oxidation	Brown-black
		Crystal	Minute flitters.
			Apply evenly.

CR 316 G224

Flint	39	Temperature	C/9
Borax	28	Surface @ C/9	Semigloss/matt
Whiting	14	Fluidity	None
Kingman feldspar	10	Stain penetration	None
Boric acid	9	Opacity	Opaque
	100	Color / oxidation	Lizard, metallic
Red iron oxide	20		plum, when thin
			Black when thick
		Crystal	Thick silver fleck

Crystal Glazes

Crystal glazes are beautiful and considered very difficult to obtain. They are not produced commercially to any significant extent, with the exception of floor and wall tiles that use rutile crystal earthenware and mid-temperature glazes. However, a few ceramists are now experimenting with crystal glazes, and interest in the use of the glazes is ever increasing.

At the Copenhagen Porcelain Works in the late nineteenth century, crystal glazes were produced by first applying a porcelain glaze to a porcelain pot, which was then fired to maturation. The porcelain glaze, applied in a medium-thick coat, was glossy, transparent, and nonflowing. Over the fired glaze a thick coating of a silica–zinc mixture was applied. The pot, now coated with the mixture, was refired to maturation temperature and slow-cooled. Zinc silicate crystals formed on the surface. Metallic oxides were added to the silica–zinc mixture for color. It was discovered, moreover, that fritting the majority of minerals used in the glaze increased both the formation and the size of the crystals.

The most successful Copenhagen glaze consisted of two frits:

Frit A		Frit B	
Flint	54.5	Zinc oxide	82.9
Zinc oxide	24.5	Flint	14.3
Pearl ash	20.9	Pearl ash	2.8
	99.9		100.0

Glaze	C/11 oxidation firing
Frit A	70
Frit B	30
	100

In both Europe and the United States during the late 1800s and early 1900s, several commercial potteries were successfully producing highly innovative ceramics. Even though the pieces were produced at a factory individual potters and decorators signed their works. Most of the commercial ceramics made during this period were Art Nouveau, Japanese-influenced styles, or asymmetrical floral patterns. A few particularly creative potteries initated the widely copied technique of employing a single glaze—such as oil spot, aventurine, sang-de-boeuf, mottled, or crystal—as the sole means of decoration. The potteries that produced crystal glazes include Universal City Pottery, Missouri (Figure 5); Fulper Pottery, New Jersey (Figure 6); Sèvres, France (Figures 7–9); Rockwood Pottery, Cincinnati (Figure 10); and the Royal Porcelain Factory, Denmark.

During this period, too, notable individual ceramists, not working for factories, were turning out handmade ceramic forms. These ceramists were responsible for the revival of individual potters in America and Europe. The spirit of individuality has continued, and contemporary potters using crystal glazes include Laura Andreson, Los Angeles (Figures 11 and 12); Rose Cabat, Tucson (Figures 13 and 14); John Kent Baker, Sunnyvale, California (Figures 15 and 16); Robert Bixler, Los Gatos, California (Figure 17); Jack Felton, Manton, California (Figures 18–22); Steven McGovney, Seattle (Figures 23 and 24); and Herb Sanders, Wichita (Figures 25 and 26).

CRYSTAL FORMATION

A crystal glaze is one which, upon cooling, forms ice-like structures on its surface. The basic crystal glaze consists of silicate glass, fluxes, and crystal formers (such as zinc for high temperatures and chrome or uranium for low temperatures). Other metallic oxides are added for coloring and glaze stabilization. Crystal formation occurs in two steps: first, the atoms arrange themselves to form a unit cell, or nucleus; second, additional groups of atoms attach themselves to the nucleus to increase its size. This network structure creates the crystal form. The growth of nuclei depends on the attraction of the proper atoms and on the rejection of unwanted atoms from the forming glaze. In order for growth to take place, furthermore, the composition of the crystal must be different from that of the glaze; the glaze composition must be suitable; temperatures must be high enough to allow the glaze sufficient mobility and fusion; and the crystal formers must be in excess—that is, the fluid glaze must have more of the crystal-forming material than it needs and thus readily give it up to the forming crystal.

It is a long, often frustrating road to control of even the simplest crystal glaze. Empirical research must be undertaken before much progress can be made. A thorough review of the literature reveals little but confusing and questionable data. It was concern about this situation that prompted much of the research that is reported in this book. The information given does not represent an attempt to define exacting chemical and physical laws in ceramics, or to analyze compounds and procedures. Rather, the results of observation, experimentation, literature researching, and discussions with the leading crystal-glaze ceramists have been capsulized in order to provide how-to-do-it techniques and formulas.

CRYSTAL GLAZE TECHNIQUES

The ideal crystal glaze would be a raw type that does not require the fritting of the materials; matures at the same temperature as the clay body; does not craze; does not flow and thus does not require the use of a catch basin or the grinding of the bottom;* does not need a special cooling cycle; can be reduced or fired with a regular stoneware reduction firing; and will produce large and evenly distributed crystals every time. The state of the art of crystal glazing is such that several of these characteristics are possible in any given glaze, but not all at the same time. With careful handling, accurate

* Because crystal glazes are fluid—some very, others only slightly—they may require a catch basin. A catch basin can be eliminated with some of the formulas if one uses a deft hand in spraying and consistent firing cycles and temperatures. If a catch basin (or equivalent device) is needed, see pages 215–21.

testing, and patience, the determined ceramist can achieve many of the qualities using the formulas given in this chapter.

It is expected, of course, that a clay body and glaze of the same maturing temperature will be used. The tendency of lower-temperature glazes to craze can be lessened if the clay body is bisque-fired at a higher temperature. Crystals will form in a regular stoneware oxidation firing and cooling cycle, but their size will generally be small. A special cooling cycle is one of the key factors in the crystal firing; fortunately, it is only a minor inconvenience. Singlefire crystal glaze is very, very difficult, and for this reason the author has not researched this approach.

CLAY BODY

It is an established fact that a given glaze will often yield different results when used on different clay bodies, in part because of the difference in absorption of the body materials into the glaze. Some clay bodies containing amounts of free alumina iron, salts, metallic oxides, fluxes, and silica are absorbed into the glaze to such an extent that the character of the crystals is altered. Most white, buff, or light gray clay bodies, on the other hand, will not interfere with the formation of the crystals. If crystals do not form at all on the clay body, then a different body should be tried.

GENERAL CONSIDERATIONS FOR HIGH-TEMPERATURE
CRYSTAL GLAZES

Many factors influence the development of crystal glazes, including clay composition; glaze composition; type of nucleus in crystal development; glaze thickness; colorants in the glaze; firing time; maturation temperature; soaking time at maturation point; character of kiln atmosphere; impurities in the kiln's atmosphere; and cooling cycle. Each crystal glaze composition has a different time-and-temperature cycle that determines the rate and size of crystal growth. General considerations for development of crystal glazes are:

1. High-soda-content glazes develop crystals more readily than do potash- or calcium-base glazes.

2. Manganese or iron in small amounts—generally under 2 percent—helps promote large crystals.
3. The proportion of zinc in the glaze formula should range from 10 to 35 percent in a zinc-silicate crystal type.
4. Titanium produces small but evenly distributed crystals.
5. The growth of crystals takes place in the cooling-soaking cycle of the firing. This must be a prolonged period to provide enough time for the crystal-forming bonds to be attracted to one another.
6. The more time allowed for crystal growth, the larger and more plentiful the crystals.
7. To ensure favorable crystal growth, the ingredients in the glaze magma must combine and fuse at the peak firing temperature.
8. Reaching the maturation temperature as quickly as possible results in better crystal formation and thicker glaze, especially on the edges.
9. Thicker glaze application on top edges prevents the fired glaze from being too thin.

FRITS

Research indicates that fritting the majority of the materials in the crystal glaze is necessary. (There are some exceptions.) Some raw and frit-and-raw mixture glazes have been successful. Still, in most crystal glazes the use of fritted materials is the only way to go.

The majority of formulas for crystal glazes call for commercial frits, or the fritting of all or part of the materials. The two commercial frits most frequently used today for crystal glazes are FERRO #3110 and PEMCO #283. Unfortunately they are in limited supply and sometimes difficult to obtain. Some large ceramic supply houses will special-order them, but only in minimum amounts of 50 to 100 pounds. Other frits that have proved successful are PEMCO #626 and #25, FERRO #3134, as well as sodium silicates SS 56 and others.

When preparing one's own frits, the following procedure is recommended. After the formula has been selected, it is made up and tested on the type of clay body that will be used in making the

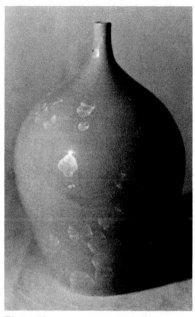

Figure 58

Figure 58 John W. Conrad, *Vase,* porcelain, crystal (CR 321) glaze, 7½″ high

Figure 59 John W. Conrad, *Vase,* porcelain, crystal (CR 322) glaze, 9″ high

Figure 60 Detail of Figure 59

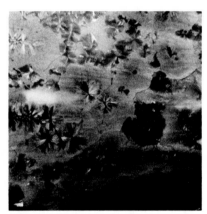

Figure 60

Figure 59

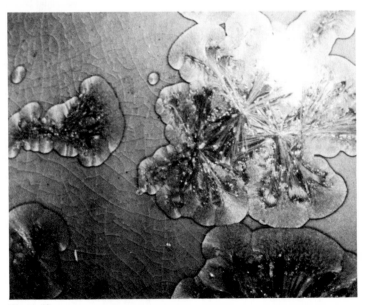

Figure 61 John W. Conrad, *Vase,* porcelain, crystal (CR 331) glaze, 10¼″ high

Figure 62 Detail of Figure 61

Figure 61

Figure 62

Figure 63

Figure 63   John W. Conrad, *Vase,* porcelain, crystal (CR 331) glaze, 7" high

Figure 64   John W. Conrad, *Vase,* porcelain, crystal (CR 331) glaze, 10¾" high

Figure 64

Figure 65

Figure 65   John W. Conrad, *Vase,* porcelain, crystal (CR 335) glaze, 10¾″ high

Figure 66   Detail of Figure 65

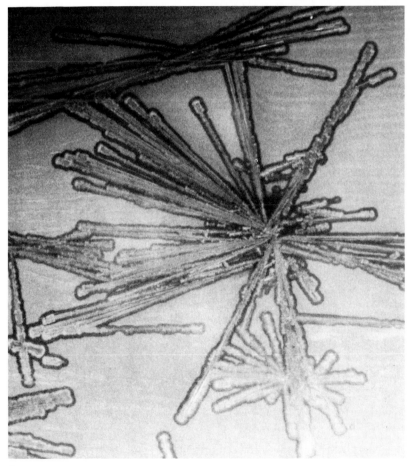

Figure 66

Figure 67 John W. Conrad, *Vase,* porcelain, crystal (CR 339) glaze, 7¾″ high

Figure 68 Detail of Figure 67

Figure 67

Figure 68

Figure 69 John W. Conrad, *Vase,* porcelain, crystal (CR 342) glaze, 9¼″ high

Figure 70 Interior View of Broken Vase (Figure 69)

Figure 69

Figure 70

Figure 71

Figure 71 John W. Conrad, *Vase,* earthenware, crystal (CR 347) glaze, 7¾″ high

Figure 72 John W. Conrad, *Vase,* earthenware, crystalline matt (CR 353) glaze, 7¾″ high

Figure 72

pots; for the test, the glaze-application methods and firing procedures should be those that would be suitable if the test were on a crystal pot. The test is done first on conventional test pieces, and the glazes that survive are then tested on small pots. Five hundred grams or more of the glaze materials are weighed, mixed, and placed in a crucible. A crucible can be made from a high-alumina fire clay formula (see Chapter II for formulas). The crucible is fired at one to two cones higher than the melting temperature of the frit.

The crucible, with mixture to be fritted, is placed in a kiln and fired to the point at which the mixture melts and the bubbles have cleared from the melt. At that time, sturdy tongs are used to remove the crucible from the kiln, and the hot glass is poured into a metal bucket filled with cold water. This is a dangerous procedure, and extra safety measures should be observed. The shattering of the liquid glass into small pieces facilitates grinding. The fractured glass is placed into a ball mill and ground for two hours. The water is changed and the grinding is continued until the mixture will pass through a 100-mesh screen. Washing the frit will remove soluble sulphur, salts, and alkalines. The frit is set to dry; after drying, it is placed in a covered, labeled container.

It should be remembered that when liquid frit is poured out of a crucible, there will be a thin coating left on the inside. This coating may contaminate the next frit, especially if it is of a different composition. The cost and effort of making one's own crucibles is such that the use of a separate crucible for each frit is desirable. Or if, as is possible, different frit compositions can be melted in the same crucible, care must be taken to prevent contamination. Another point to consider is that the time and peak temperature of the melt will determine the amount of alumina extracted from the crucible by the frit. In some crystal formulas the tolerance for alumina is extremely small, and if the alumina content goes over the tolerance level it will contaminate the glaze and inhibit the growth of the crystals.

When testing the glaze, make only enough frit to be used immediately. Should the test prove successful, then larger amounts can be made. When a good frit is discovered, at least five pounds or more should be made, for there is a fair amount of work involved in making frits and one does not want to repeat the process more than is necessary. Even though making one's own frits is somewhat difficult, it can add greatly to the satisfaction the ceramist derives from crystal glazing, for he or she is thus involved in the total process of creating a crystal glaze from beginning to end.

SEEDING AGENTS

To *seed* the glaze is to ensure that crystals will form, sometimes in particular, distinct places on the pot. Granular rutile, ilmenite, vanadium, titanium, molybdenum, zinc oxide, and zinc silicate are examples of seeding agents. There are several ways to employ these agents: (1) They can be put in the liquid glaze just before it is sprayed on the pot to produce a random crystal pattern. (2) They can be blown onto the hot glaze through the kiln spy hole, using a metal tube. (3) Each particle can be glued (with CMC) to the desired spot on the bisqueware and the glaze sprayed over it; or the particles can be placed on top of the sprayed glaze. The placement can be random, in bands, in prepared designs—or one can even form one's own name in crystals. (4) The pot can be fired with crystal glaze in a regular stoneware firing. Afterwards, the grains are glued (with CMC) on the glaze, and the pot refired in a crystal firing cycle.

Crystal configurations produced by these methods are more controlled and consistent than are the patterns that nonseeded methods yield. Seeding is particularly necessary for those crystal formulas in which it is difficult to get the nuclei started. The seeding material need not be the same as the material in the crystal glaze formula; it merely acts as a nucleus.

CRYSTAL ENHANCERS

Crystal enhancers are used to speed up the formation of crystals; to change their pattern, shape, or growth; to increase their number; or to reduce the amount of soaking time needed to complete their formation. The enhancers work in somewhat the same way seeding agents do, but enhancers can affect the shape of the crystal, whereas seeding agents cannot. Small amounts—usually under 1 percent— of iron, manganese, cadmium, molybdenum, vanadium pentoxide, and/or tungsten are added to the glaze formula. Molybdenum, cadmium, and tungsten can by themselves form crystals, but not dependably or easily. The best crystals are formed by using these in combination with zinc. Iron and manganese will give color to the glaze if used in amounts over 2 percent.

MODIFYING THE FIRED CRYSTAL-GLAZED POT

The appearance of an already-fired pot can be altered or further enhanced by various methods. The well-known ceramist John Kent Baker, who has the notable success rate of 60 percent in his first firings, has stated that the successful pots, as well as the unsatisfactory ones, can be significantly improved through modification techniques. Some of the most effective of these techniques are the following:

1. A pot that has few or no crystals, bubble imperfections, or thin glaze areas should be refired. The pot is reglazed—entirely or only over defective areas—and then refired. Refiring will often produce crystals that did not occur in the first firing.

2. Formation of crystals is ensured by the application of a thin layer of silica (30 percent) and zinc (70 percent) over the first glaze, or by the seeding of the surface of the glaze (see "Seeding"). The pot is then refired in a crystal firing cycle. Double-firing with a silica–zinc mixture is one of the oldest methods of obtaining crystals; it was first used in Copenhagen in the 1880s.

3. Some crystal glazes will craze, and their appearance can be enhanced by rubbing india ink into the cracks. Another approach is to rub a metallic oxide into the crack, wipe off the excess, and refire the pot. The oxide will fuse into the glaze, which will retain the crackle pattern but form new cracks when the pot cools. A red india ink can then be rubbed into the new cracks to produce a two-color crackle pattern.

4. All the crystal glazes are meant to be neutral or oxidation fired, but under certain conditions they can be reduction fired. The crystals generally will not form when the surface is under the tension that a reduction atmosphere produces. *Sometimes* the fired pot can be refired in a lower-temperature stoneware reduction firing, which will drastically change or alter some colors without destroying the crystal pattern. In some cases, too, the copper greens will become brilliant reds or take on a copper luster. A second method is to

reduce the crystal firing load during the last one-half hour of the soaking period. This will not interfere with the crystals that have already formed. There is a third method: when the kiln has cooled to 1400°F, the kiln atmosphere is heavily reduced for fifteen minutes; then the dampers are kept closed as the kiln cools. While the glaze is still slightly molten, the active carbon in the atmosphere will affect the iron and copper colorants in the same way it does in a raku firing.

COLORANTS FOR CRYSTAL GLAZES

Metallic oxides and stain colorants for crystal glazes are essentially the same as for any other type of glaze. The only notable difference is that crystal glazes may have two colors: that of the background and that of the crystal. The glaze and the colorant must be well mixed either by mechanical mixers, mortar and pestle, or a ball mill, and then sieved through a 60-mesh screen. Otherwise, the glaze will not be uniformly combined, and an undesirable speckling of the glaze may result. Most poorly mixed glazes, in fact, will not even produce crystals. A note of exception: Very successful crystal glazes have also been produced by putting the weighed materials into a blender, mixing, and then applying the *unscreened* mixture to the pot. In other words, even the materials that would not pass through a 60-mesh screen are used. This method works even with uranium, which is composed of hard lumps and usually requires several hours of ball-milling to grind it small enough to pass through an 80-mesh screen.

The colors available with metallic oxides in crystal glazes are similar to those obtainable in regular stoneware glazes. Chrome oxide at low temperature will produce orange, orange-red, and reds, while at high temperature it will give the classic chrome green. Cobalt at high temperature will, under certain conditions, produce pink instead of the traditional blue. Copper will produce blue, blue-green, or green colors, depending upon the composition of the glaze. The colors that are possible using a C/10 raw crystal glaze, as reported by Thiemecker, are as indicated below.*

* H. Thiemecker, "Notes on Cone 10 Raw Crystal Glazes," *Journal of American Ceramic Society*, No. 17 (1934), pp. 359–62.

Table 6
C/10 Raw Crystal Glaze Results

Colorant	Percent	Color of Glass Matrix		Color of Crystal	
		Min. added	*Max. added*	*Min. added*	*Max. added*
Manganese	2.0–10.0	Light brown	Dark brown	Cream	Flesh
Iron oxide	2.0–10.0	Greenish-brown	Dark brown	Fawn	Light brown
Chrome oxide	0.5– 2.5	Pale green	Apple green	Emerald green	Emerald green
Soda uranium	2.0–10.0	Yellow-green	Yellow-green	Canary yellow	Gray
Copper oxide	1.0– 5.0	Apple green	Dark green	White	White
Cobalt	0.5– 4.0	Powder blue	Navy blue	Pink	Pink
Ceramic black	2.0–10.0	Blue-black	Blue-black	Fawn	Light brown

An extensive testing series on a crystal glaze to which various colorants were added was conducted by H. M. Kramer, who used the following base glaze*:

Sodium silicate, dry	52.3
Silica	20.5
Zinc oxide	18.9
Titanium oxide	8.4
	100.1
Water	5.1 (+/−)

The colorants were added to the glaze and ball-minded. The mixture was applied to bisque-fired vases by painting in one or two coats. The firings took ten to twelve hours to reach maturation point at C/9 half-down. The temperature was then dropped 200°F and held constant for three hours, after which the kiln was allowed to cool at its own rate. Table 7 is a condensed version of the original study. The color of both the glaze and the crystals will be different with other formulas; Kramer's results provide an indication of the colors possible.

* H. M. Kramer, "Colors in a Zinc Silicate Glaze," *Journal of American Ceramic Society,* No. 7 (1924), pp. 868–77.

Table 7
Colors for Crystal Glazes

Oxide and percent							Crystal color	Glaze color
Cobalt carbonate	Manganese carbonate	Copper carbonate	Uranium oxide	Nickel carbonate	Iron oxide	Chrome oxide		
0.01							Pale blue	Creamy
0.05							Dark blue	Buff
0.10							Dark blue	Brown
	0.03						Fawn	Buff
	0.05						Fawn	Brown
	0.10						Dark fawn	Dark brown
		0.01					Very pale green	Pink and cream
		0.05					Copper green	Pink to tan
		0.10					Dark green	Brown-tan
			0.02				Slight creamy	Creamy
			0.05				Pale yellow	Yellow
				0.01			Pale bluish-green	Yellow
				0.02			Pale bluish-green	Apple green
				0.05			Apple green	Apple green
					0.03		Colorless	Buff
					0.05		Buff	Buff-brown

1	2	3	4	5	6	7		
0.012							Pale blue	Light brown
0.012	0.012						Moderate blue	Creamy
0.012	0.012	0.012					Light blue	Yellow
							Pale green	Fawn
0.008	0.012	0.012	0.012				Bluish-green	Cream
0.008	0.012	0.012	0.012				Pale green	Pink
0.008	0.008	0.012	0.012	0.012	0.012		Pale greenish-yellow	Yellow
0.008	0.008						Pale bluish-green	Pink
0.008	0.008	0.008					Pale blue	Light buff
							Pale blue	Yellow
0.008	0.008	0.008	0.008			0.008	Green	Brown
0.008	0.008	0.008	0.008			0.008	Mossy bluish-gray	Greenish-brown
0.008	0.008	0.008					Pale blue	Yellow
0.008	0.008	0.008	0.008			0.008	Dark green	Shiny apple green
0.008	0.008	0.008	0.008			0.008	Mossy brown	Reddish-brown
0.05	0.05	0.05	0.008	0.008	0.008		Pale green	Yellow
0.05	0.05	0.05	0.008	0.008	0.008		Bluish-green	Brown
			0.008	0.008	0.008		Light blue	Pink
0.05	0.05	0.05					Dark blue	Dark brown
0.05	0.05	0.05					Very dark blue	Light brown
							Medium green	Dark brown taupe
0.033	0.033	0.05	0.05				Mossy gray	Goldstone
0.033	0.033	0.05	0.05				Pea green	Green
0.033	0.033	0.033	0.033				Dark blue	Brown
0.033	0.033	0.033	0.033	0.033			Dark blue	Goldstone
							Blue	Reddish-brown
0.033	0.033	0.033	0.033				Blue	Gray
	0.033	0.033	0.033				Green	Brown
							Blue-gray	Goldstone
	0.033	0.033	0.033				Olive green	Dark green

FLUID GLAZE AND THE POT

The design of the pot and the amount of glaze applied is a major consideration in the crystal-glazing process. Extensive overhangs, poor design, or too much glaze will cause the glaze to pool, puddle, drip onto the kiln shelf, or crack the pot (Drawing 7). The fluidity of the glaze and the beauty of the finished ware both call for simple ceramic forms.

Aventurine, crystal, and crystalline matt glazes range from non-flowing to very fluid. In crystal glazes, the very nature of the magma requires a fluid state in which the crystals can grow; most crystal glazes, therefore, have at least some flow. With experience, glazing technique can be mastered to the point at which just enough glaze is applied to promote crystal formation and to cover the pot, but not so much that the glaze runs down the pot and onto the kiln

Drawing 7
Fluid Crystal Glazes

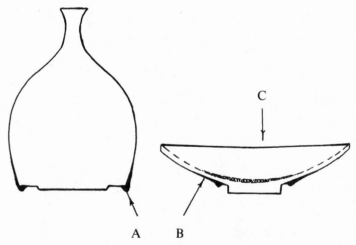

A. Fluid glaze forms heavy drip on corners and angles.
B. A ring of heavy glaze or drips develops on underside of shallow bowls.
C. The fluid glaze flows down the sides of bowls, forming a thick glaze on the bottom. If the glaze is too thick, it could cause the pot to crack.

shelf. Some glazes, on the other hand, *require* the superfluid state to develop the large crystals. Such fluid glazes will always run, so if the design of the pot does not take into consideration the flow of the glaze, the results can be disastrous. The glaze will fuse the pot to the kiln shelf; the removal of the excess glaze from the pot and the shelf is time-consuming, exasperating, and will produce a pot with a sloppy-looking bottom. A catch basin or some other means of preventing the excess glaze from pooling at the bottom of the pot is highly recommended.

CATCH-BASIN DEVICES

The gathering and control of the flow of excess glaze will necessitate both special attention to the design of the ceramic form and the devising of some means to prevent the excess glaze from pooling at the bottom of the pot or on the kiln shelf. Various methods are possible to elevate the pot and/or to gather the excess glaze. Three of the possible methods are illustrated here.

Insulation brick as catch basin. After the pot is bisque-fired, it is placed on a block of soft insulation brick (2600°F), and a line is drawn around the base onto the brick. A hacksaw or similar blade is used to cut out a square or cylindrical shape slightly larger than the base of the pot. Several coats of kiln wash are applied to the top of this block (Drawing 8). The pot is glazed and placed on top of the block for firing. During the firing the excess glaze will flow down and run off the pot onto the soft brick, which will absorb the glaze and thus prevent a heavy glaze buildup on the bottom of the pot. The soft brick with kiln wash separates easily from the pot.

Incised foot and catch basin. When the bottom of the pot is being trimmed, a line is cut with a needle through the foot from the inside towards the outside (Drawing 9). During the cutting, the clay will bulge slightly when the needle is nearly through, indicating that the cut is deep enough. A catch basin is thrown out of any clay that is handy. Both pot and catch basin are bisque-fired. The pot is then glazed, placed upon the catch basin, and fired. The glaze will flow down the pot into the basin, fusing the basin to the foot. By carefully tapping the cut line with a chisel, the pot can be broken

Drawing 8
Insulation Brick Cut as Catch Basin

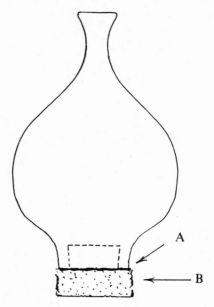

A. Several coats of kiln wash
B. Soft insulation brick (2600° F) cut into a square or cylindrical form just big enough to support the pot

away from the basin, leaving only a clean line that will require a minimum of grinding to smooth out the bottom.

Catch basin. A third method is to throw a catch basin with a cylinder in the center whose width corresponds to that of the foot of the pot (Drawing 10). The pot and the basin are thrown at the same time with the same clay. The foot of the pot is trimmed exactly to match the diameter of the catch basin cylinder. This takes planning to visualize the final, trimmed shape of the pot. The pot and the basin are bisque-fired together. Afterward, the cylinder and the foot are sanded to provide smooth, flat surfaces that match each other.

Drawing 9
Pot with Incised Line, Placed on Thrown Catch Basin

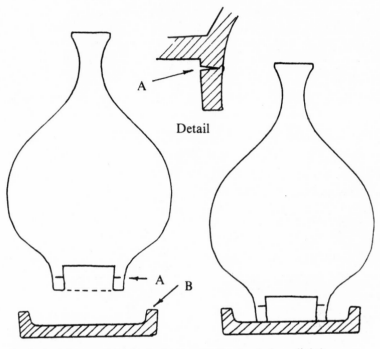

A. Incised line cut into foot of pot until the clay bulges slightly
B. Catch basin large enough to hold pot

A mixture of white glue, alumina hydrate, and a little water is painted thinly on both the cylinder and the foot, which are then joined. The pot is glazed and then fired. The excess glaze will run down the sides of the pot into the catch basin, but the alumina prevents the glaze from seeping into the seam. Light tapping with a chisel will free the pot from the basin. Very little grinding will be needed to leave a clean foot. Details and photographs of pot and catch basin preparation appear in Figures 27–55.

Drawing 10
Catch Basin with Cylinder Corresponding to Foot of Pot

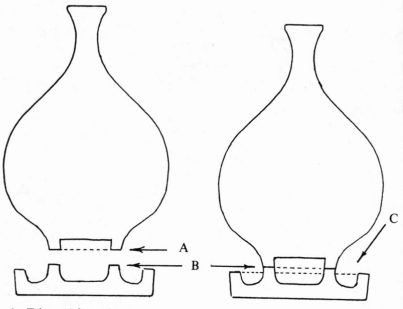

A. Trimmed foot of pot
B. Cylinder thrown to match the trimmed foot of pot
C. Catch basin with cylinder glued (mixture of white glue, alumina hydrate, and a little water) to pot

BOTTLE FORMS AND CRYSTAL GLAZE

This section illustrates in detail the step-by-step technique of shaping a bottle and of applying crystal glaze to the ware. The finished piece is a perfect example of the way a simple, classic form and a crystal glaze can enhance each other. The slender-necked bottle is one of the more difficult forms to throw. A ball of white clay—porcelain or white stoneware—is wedged and placed on the wheel (Figure 27). The use of a bat made out of masonite, plaster, particle board, or plywood is highly recommended for bottles with very slender and wobbly necks. The bat will make the removal of the bottle easier and reduce the chance of its moving off-center. Placement pins on the wheel head, with corresponding holes in the bats, make the accurate repositioning on the wheel possible, as illustrated in Figure 46.

The ball of clay is worked up and down several times until it is centered in a dome shape (Figure 28). By applying even pressure, the dome is opened up and the first pull of the wall is completed (Figure 29). The wall and the width of the bottom and the bottom sides are now established. A series of pulls raises the wall, making it thinner and straighter, gradually bringing it into a slender cylindrical form (Figure 30). The pulling continues until the cylinder is complete. By now, the bottom one-third of the bottle has been shaped. Pressure from the hand inside against the wall forces the cylinder to expand and take shape (Figure 31). The pressure exerted by the hand is repeated until the bottle takes form (Figure 32). The clay may be soft; if so, the potter must wait until it has stiffened before completing the next step. A heat lamp can be used to speed up the drying. The top two-thirds of the bottle is now pulled up and shaped. With the pressure of two fingers each on the inside and the outside, the cylinder should be forced up and in as the clay is compressed into a smaller diameter (Figure 33). To decrease further the diameter of the wall, four fingers pressing inward raise it (Figure 34). Every so often, the irregular top edge is cut off with a needle tool and a steady hand (Figure 35). As the diameter of the cylinder becomes narrower and narrower, only one finger will fit inside. The fingers on the outside continue to raise and thin the wall (Figure 36). Should the diameter of the neck become too narrow to accommodate a finger, then a pencil, dial rod, or other slender object should be inserted into the opening to give support to the outside finger pressure (Figure 37). Pressure is continued until the desired height and width of the neck is achieved. The lip can be flared and sponged to give a smooth-finished edge (Figure 38). Irregular lines, low spots, high spots, and rough areas are smoothed out with a rubber kidney-shaped tool, which is gently pressed against the wall while the bottle is turning (Figure 39).

While the bottle is still wet and soft, the clay must be thicker than normal at the bottom to give support to the form. After the clay has stiffened somewhat and can support itself, the excess at the bottom is trimmed. The catch basin is thrown at the same time as the bottle; the potter should remember that the basin must correspond to the *ultimate* size of the bottle bottom, and thus should estimate the size the bottle will be after trimming. This estimated shape may be measured with calipers (Figure 40) to determine what size the catch basin should be.

To make the basin, a small ball of clay (the same kind as is used for the bottle) is wedged and placed on a bat (Figure 41). The ball is centered in a pancake form (Figure 42). The center is opened up and pulled out to the diameter indicated by the caliper measurement (Figure 43), and then the basin's lip is turned up (Figure 44) and smoothed.

The bottle is set aside to stiffen, then returned to the wheel so that the excess clay can be removed from the bottom (Figure 45). A loop tool and scraper are used for this procedure. When trimming the bottom, the potter must keep in mind how the contour of the inside relates to that of the outside and avoid cutting too deeply. Removing most of the excess clay while the bottle is in an upright position is easier and quicker than using a trimming chuck later. The cylinder of the basin is remeasured, for it will have shrunk after drying (Figure 46). The new measurement is used to determine the exact bottom diameter of the trimmed bottle (Figure 47). After the bottle has been trimmed to size, the foot is then finished by hand trimming, or set in a trimming chuck and cut.

Both bottle and catch basin are bisque-fired. The bottom of the bottle and the top of the basin (Figure 48) are then sanded with fine sandpaper to smooth out and level their surfaces. This ensures that they will join with a tight fit and so prevent any glaze from entering the seam during firing. A thin coat of alumina hydrate, with white glue and a small amount of water, is painted on the bottle and basin surfaces (Figure 49). The two pieces are then pressed together, forcing out any excess glue, which is wiped off (Figure 50).

Only the glaze that is to be used immediately is weighed, screened, ground if necessary, and sprayed onto the form (Figure 51). The glazed bottle with glued-on basin is placed in the kiln. In this case, the bottle is placed in a test kiln with several test pieces (Figure

52). The test kiln used is gas-fired and has a 14-inch cube chamber and is lined with 5½-inch-thick fiberfax. The kiln is fired to C/9 in four hours. Although it may be able to reach high temperatures faster, the silicon carbide shelves cannot take the thermal shock. The glue will burn away in the firing, but the bottle will still be attached to the basin by the excess glaze which has run down the bottle into the basin, fusing the two together (Figure 53). To free the bottle, a light tap is made with a chisel where the bottom rests on the basin (Figure 54). Figure 55 shows that just a thin layer of glaze holds the two together. The bottom of the bottle will be smooth and level, and only a light grinding is necessary to remove any sharp edge. The finished bottle and a close-up of the crystal pattern (glaze CR 343) are pictured in Figures 56 and 57.

GLAZE APPLICATION

Spraying is the recommended method of applying crystal glaze to a pot. The glaze should be applied to about the thickness of a dime, with the very top, lip, or rims slightly thicker. A dipping technique, if used, should have the glaze at 1.6 specific gravity on bisqueware. A quick dip will be below 1.5, which is too thin a coating. Above 1.7, without a deflocculant, the coating is too thick and will flake off in the firing or when dry. Since the glaze is applied very heavily, crawling can be a problem. With a painting technique, the glaze is applied in several thin layers rather than one heavy one.

Two hundred grams of dry glaze is needed to cover one average ten-inch bottle; four hundred grams is needed for a fourteen-inch bottle. Only the amount of glaze needed for the glaze session is mixed. Any remaining glaze is discarded or put in a scrap glaze container. This scrap can produce interesting, unexpected, and unduplicatable glazes. Discarding may seem wasteful, but better, more predictable results are obtained by using only fresh glaze. Many raw-glaze formulas that contain soda ash, pearl ash, sodium bicarbonate, borax, or boric acid will cake in the liquid glaze, sometimes in less than one hour. A binder, or adhering agent, is added to the glaze mix. Binders are CMC, gum tragacanth, gum arabic, egg whites, bentonite, and ball clay.

The inside of the pot is glazed first. It is not necessary to use a crystal glaze for this. Any glaze that matures at the same temperature is acceptable; however, the flow of the crystal glaze can be an attractive asset to the inside of a pot. Ceramic forms with visible insides, such as bowls and wide-mouth bottles, are covered on the inside with a thin coating of a nonflowing glaze whose color will complement the crystal glaze. Over the top one-eighth to one-fourth of the glazed inside area, a thin coating of crystal glaze is applied by spraying, dipping, or painting. Care must be taken with painting so that the under-glaze is not lifted, nor too much glaze applied. The ware is fired in a regular stoneware or crystal firing. The crystal glaze will flow over as much as three-fourths to seven-eighths of the inside of the pot. The effect is very striking, with or without crystals. The stoneware firing does not produce large crystals, but the two-glaze combination is an attractive blend of under-glaze and crystal glaze that creates mottled, streaked, or dual-colored patterns.

More than one crystal glaze can be used on the same pot. The potter can use two or more batches of the same crystal glaze, each having a different colorant, or two or more different glaze formulas. The glazes are applied on different parts of the pot, with or without extensive overlapping. The glaze is sprayed in wide bands or stripes, patches, and irregular patterns; the potter must be sure that the entire pot is covered with a thick glaze. Because of the glaze flow and the crystal growth pattern, detailed glaze application designs will not show.

FIRING CRYSTAL GLAZES AT CONE/4 AND HIGHER TEMPERATURES

The composition of the glaze is the most important element of crystal glazing, with firing a close second. The following firing schedule is recommended for cone 4 and higher crystal glazes, but it will need to be adjusted according to kiln, glaze, and fuel used.

1. The kiln is heated slowly to 600°F, to drive out both the physical water and most of the chemical water. This process takes from one to three hours.
2. From 600°F up, the kiln is fired in an oxidative atmosphere as rapidly as fuel type, kiln-shelf composition, and kiln design permit. The firing is done at the rate of 200°F or more

per hour until the kiln reaches the maturing range of the glaze.

3. The peak or maturing temperature is maintained (soaking period) for at least ten minutes, up to one hour for some glazes. This step permits the gas to escape, craters and pinholes to heal, the ingredients in the glaze to mix thoroughly, and the crystals to begin developing.

4. The damper openings are closed somewhat, the amount of fuel is decreased sufficiently to allow the kiln to cool down slowly for one hour until its temperature has dropped 200°F. The rate of cooling will have an influence on both the size and the number of crystals produced.

5. The dampers and fuel are adjusted again to maintain the lowered temperature for a period of one to six hours. The length of time this temperature is held will depend, as is so often the case, upon kiln design, fuel type, crystal formula, and the crystal size desired. In general, the longer the soaking period, the larger the crystal. However, the maximum size the crystal can attain is determined, fundamentally, by the glaze formula. Usually the soaking time will average about three hours.

6. Halos, or growth rings (similar to the growth rings of a tree) around crystals, are produced by raising and lowering the kiln temperature several times during the soaking period.

7. It is possible to change some glaze colors by introducing a reductive atmosphere in the kiln during the last half hour of the soaking period. By this time the crystals have already formed, so the reduction will not interfere with their formation.

8. At the end of the soaking period the fuel is shut off and the kiln is allowed to cool at its own rate.

9. Some ceramists have experimented with the reduction of the kiln at 1400°F (+/−) to obtain metallic-like surfaces, crystals, or color changes.

The firing must be stopped at the maturing temperature stage so that sufficient glaze will remain on the pot to form the crystals. Should the glaze have a strong tendency to form crystals, then its soaking period need not be very long. Conversely, those glazes that tend to form crystals slowly must be given a longer soaking period. The duration of the soaking period is an important consideration. A soaking period which is too long will cause the crystals to spread out and cover the surface entirely, at which point the individual crystals become indistinguishable. Firing time must be carefully watched too, for if the glaze is overfired or is kept at peak temperature for too long, the glaze will run off and leave a sandy layer too thin for crystals to form.

TEMPERATURE CONTROL

Electric kilns are the easiest to control in both the firing and the cooling cycles of crystal firing. The preheating, drying, and firing-to-maturation-temperature procedures are the same as those for any glaze firing. It is in the cooling cycle that technique is critical. Here, precise temperature control is vital, and an accurate pyrometer is an essential part of the kiln equipment. Pyrometic cones are adequate for indicating the maturing temperature, but they are useless for temperature indication during the cooling cycles. In the cooling cycle the temperature in either electric or gas kilns can be manually controlled. The kiln operator monitors the kiln temperature as registered on the pyrometer and adjusts it by hand according to the time and temperature requirements of the glaze. Various temperature-regulation equipment is available to control both firing and cooling cycles. Some are programmable and automatic, while others will simply hold or shut off the kiln at a set temperature. The following describes several different temperature-regulating devices:

An *automatic shut-off pyrometer* is a temperature indicator with an adjustable shut-off control that enables the operator to choose the temperature at which the kiln will go off automatically. The cooling cycle is controlled by hand.

A *temperature-holding pyrometer* operates much like the thermostat in a hot-water heater, electric skillet, or oven. The pyrometer will maintain the temperature set by the operator. Most pyrometers of this type have a dual-control feature which regulates both maintenance of temperature and temperature shut-off. Temperature-holding pyrometers can turn off the kiln at maturing temperature and maintain the soaking temperature until the operator shuts the kiln off manually. The operator does not need to be at the kiln during the soaking period, as with the automatic shut-off type.

A *mechanical time-and-temperature programmer control* operates on a twenty-four-hour clock or similar device. A *cam drive* is a piece of sheet metal cut to a pattern (Drawing 11) that corresponds

Drawing 11
Cam Drive Time-and-Temperature-Controlling Device

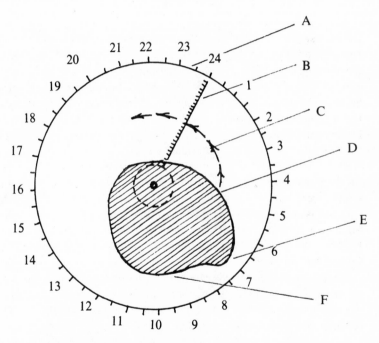

A. Time positions on disc drive
B. Temperature scale
C. Direction of cam travel
D. Cam drive: sheet metal cut to shape according to firing schedule
E. Notch for maturing temperature and soaking period
F. Soaking period during the cooling cycle

to a firing schedule and is placed in front of the clock face. As the cam revolves, the edge of the sheet metal sets off the various electric switches which will regulate the firing schedule. A *disc drive* works in much the same way as the cam drive, except that instead of sheet steel cut to a pattern it uses a metal disc that contains hundreds of holes. The operator places metal pegs into the holes that correspond to the temperature and time cycle desired. With the cam drive, a separate pattern must be cut for each different time and temperature setting; but with the disc drive, the pegs can be shifted to various positions on the same disc. Both drives have a manual override to adjust the firing, enabling the operator to compensate for load and other differences in kiln firing.

An *electronic controller* is a device on which the operator adjusts one or more dials or buttons on a unit to control both the time and temperature of the firing or cooling cycle. Some models have dual controls that can regulate both the firing and cooling cycles. These controls can be reset any time during the firing or cooling cycles.

FUMING

Fuming is a process that gives a glazed pot a mother-of-pearl, or lustrous, finish. The technique works with earthenware, stoneware, porcelain, and salt- or crystal-glazed ware. The pots are first fired; then, in the cooling cycle, when the kiln contents have reached dull-red temperatures, a fuming agent—either stannic chloride, bismuth subnitrate, or zinc chlorate—is placed in the kiln with a long-handled spoon. Three to five spoonfuls are inserted through the spy holes, burner ports, fire box, or other openings. About 2 ounces will suffice for a 30-cubic-foot kiln. At the dull-red temperature level the chemicals will turn into a vapor, fuming an extremely thin coating over the glaze surface and everything else in the kiln. Note: *The vapors are corrosive and in a closed room could become highly toxic.* Stannic chloride produces a transparent mother-of-pearl finish; bismuth produces a lustrous effect; and zinc can produce greenish, yellowish, or grayish effects. While stannic chloride is the most popular and provides the best results, a mixture of 90% stannic chloride with 10 percent bismuth and/or zinc will also give attractive results.

HIGH-TEMPERATURE CRYSTAL GLAZE FORMULAS

The formulas given here are for raw, fritted, or mixtures of raw and fritted ingredients. The formulas that call for partially or entirely fritted materials will produce the highest percentage of successful crystal glazes. All the formulas given will produce crystals. In the testing, the soaking ranged from periods as short as one hour to as long as five hours. Where the formula indicates that the crystals were small, then most likely the soaking period was relatively short, perhaps an hour or so. The same glaze, if soaked for several hours longer, could produce large crystals. The crystal designs range from a galvanized-iron effect, to individual stars and snowflakes, to needle shapes, criss-cross patterns, or broken surfaces. All the formulas will need to be tested by the potter to adapt them to the kiln, clay body, and firing techniques. Because of the complicated nature of crystal-glaze preparation and firing, it is not likely that every formula will work on the very first try.

Examples of crystal-glazed pots and close-ups of individual crystals are found in Figures 58 through 70.

CR 317 Over-glaze Crystal
(To be used over a glazed surface)

Sodium silicate SS 56		Temperature	C/8–11
(dry)	60.0	Surface @ C/9	Semigloss
Zinc oxide	39.9	Fluidity	Little
Ilmenite	0.1	Stain penetration.	All
	100.0	Opacity	Transparent
		Color / oxidation	Slight white
		Crystal	Small, all-over crystals

CR 318 Frit #626 Base

Frit #626 PEMCO	60.0	Temperature	C/9
Zinc oxide	22.0	Surface @ C/9	Gloss (2-hour soak)
Silica	15.0	Fluidity	Some
Molybdenum oxide	3.0	Stain penetration	All
Ilmenite	0.1	Opacity	Transparent
	100.1	Color / oxidation	Window clear
		Crystal	⅜" white crystals

CR 319 Oxford Feldspar

Oxford feldspar	60	Temperature	C/9
Zinc oxide	16	Surface @ C/9	High gloss (2-hour
Barium carbonate	11		soak)
Whiting	8	Fluidity	Fluid
Flint	5	Stain penetration	All
	100	Opacity	Transparent
		Color / oxidation	Water clear
		Crystal	Small white
Ilmenite	0.1 ⎤	Color / oxidation	Amber (34)
Nickel carbonate	4.0 ⎦	Crystal	Blue crystals (83)
		Note: Cracks where thick	

CR 320 Jack Pott's Crystal Brown

Frit #283	59.9	Temperature	C/8
Zinc oxide	12.2	Surface @ C/8	Gloss (no soaking
Silica	6.5		period)
Manganese dioxide	5.7	Fluidity	Nonflowing
Yellow ochre	4.1	Stain penetration	Darks only
Whiting	4.1	Opacity	Transparent
Rutile	2.0	Color / oxidation	Medium brown
Dolomite	1.6	Crystal	⅜″ flower shape, gray
Bentonite	1.6		
Nepheline syenite	1.6		
Alumina hydrate	0.8		
	100.1		

CR 321 Snowflake Crystals

Frit #3110 FERRO	57	Temperature	C/9
Zinc oxide	25	Surface @ C/9	Gloss (3-hour soak)
Flint	18	Fluidity	Fluid
	100	Stain penetration	All
		Opacity	Transparent
		Color / oxidation	Window clear
		Crystal	Snowflake, frost white
		Note: Cracks where thick	

Colorant		Color of Glaze	Crystal
Cobalt	1.2	Medium blue (93)	Medium blue (93)
Copper carbonate	3.0	Light green (133)	Light green (133)
Manganese carbonate	3.0	Light pink (7)	Light pink (7)
Red iron oxide	2.0	Off-white (8)	Off-white (8)
Nickel carbonate	4.0	Amber (10)	Blue (98)

CR 322 Frit #3124 Base

Frit #3124 FERRO	56	Temperature	C/9–10
Zinc oxide	37	Surface @ C/9	Gloss
Flint	7	Fluidity	Some
	100	Stain penetration	All
		Opacity	Transparent
		Color / oxidation	Clear
		Crystal	White crystals

Colorant		Color of Glaze	Crystal
Nickel carbonate	4	Cream (29)	Medium blue (85)
Iron oxide	5	Warm gray-tan (148)	Darker gray-tan (157)
Cobalt carbonate	2	Blue (91)	Blue (91)
Copper carbonate	4	Light sea green	Light gray-green (139)
Chrome oxide	2	Mauve (dark 46)	Grayed pink
Black uranium oxide	8	Grayed yellow-green (150)	Grayed yellow-green (150)
Cobalt carbonate	1.5 ⎫	Golds	Light blue and gold
Black uranium oxide	6.0 ⎭		

CR 323 Indiana University

Frit #3124 FERRO	55	Temperature	C/9–10
Zinc oxide	35	Surface @ C/9	Gloss
Flint	6	Fluidity	Fluid
Titanium dioxide	4	Stain penetration	Most
	100	Opacity	Translucent
		Color / oxidation	White
		Crystal	Large cream-white crystals

CR 324 Galvanized-Metal-Effect All-Over Crystals

Frit #25 PEMCO	55	Temperature	C/9
Zinc	30	Surface @ C/9	Fine sandpaper matt
Silica	15	Fluidity	Little
	100	Stain penetration	All
		Opacity	Translucent
		Color / oxidation	Slight white
		Crystal	Galvanized-metal-effect all-over crystals

CR 325 Frost-Like Crystals

Frit #25 PEMCO	55	Temperature	C/9
Zinc oxide	24	Surface @ C/9	Gloss
Silica	15	Fluidity	Fluid
Barium carbonate	6	Stain penetration	All
	100	Opacity	Transparent
		Color / oxidation	Clear
		Crystal	Frost-like crystals

Colorant		Color of Glaze	Crystal
Uranium	8.0	Transparent yellow (14)	Gold (14)
Nickel carbonate	4.0	Amber (10)	Blue (99)
Copper carbonate	4.0	Green (130)	Green-gray
Cobalt carbonate	1.2	Blue (93)	Blue (93)
Copper carbonate	4.0 }	Medium blue-green	Blue (99)
Cobalt carbonate	0.3 }		

CR 326 #103

Frit #3124 FERRO	53	Temperature	C/9
Zinc oxide	33	Surface @ C/9	Gloss
Flint	6	Fluidity	Fluid
Titanium dioxide	3	Stain penetration	Darks
Tin oxide	3	Opacity	Opaque
China clay	2	Color / oxidation	Cool white (7)
	100	Crystal	White
Nickel	4	Color / oxidation	Sea green (126)
		Crystal	Darker sea green (126)

CR 327 Zinc-Titanium Crystals

Frit #3124 FERRO	52.9	Temperature	C/9–10
Zinc oxide	33.0	Surface @ C/9	Gloss
Flint	7.0	Fluidity	Some
Titanium dioxide	7.0	Stain penetration	All
Ilmenite	0.1	Opacity	Translucent
	100.0	Color / oxidation	Light gray-blue
		Crystal	Large light gray-blue

CR 328 Laura Andreson's Crystal Base

Kingman feldspar (old		Temperature	C/9
Kingman)*	52.7	Surface @ C/9	Matt
Colemanite	12.5	Fluidity	Little
Barium carbonate	11.4	Stain penetration	Little
Whiting	7.8	Opacity	Opaque
Titanium oxide	7.0	Color / oxidation	Soft white
Zinc oxide	5.3	Crystal	Subtle broken pattern
Kentucky old mine #4	3.3		
	100.0		
Copper carbonate	1.0	Color / oxidation	Soft gray-green
		Crystal	Light gray crystals

CR 329 Cornwall Stone

Cornwall stone	52.5	Temperature	C/9–10
Zinc oxide	17.0	Surface @ C/9	Gloss
Silica	15.1	Fluidity	Little
Whiting	11.2	Stain penetration	Most
Georgia kaolin	4.2	Opacity	Opaque
	100.0	Color / oxidation	Whitish
		Crystal	Small white crystals

* Kingman feldspar that was mined before 1976 had less than 1 percent iron oxide in the mineral. The Kingman mined today has 6 percent iron. This difference will greatly affect crystal and many other glazes, clay bodies, engobes, etc., but not dark-colored glazes and clay bodies.

CR 330 Transparent Crystals

Frit #3110 FERRO	50	Temperature	C/9–10
Zinc oxide	28	Surface @ C/9	Semigloss (2-hour soak)
Flint	22	Fluidity	Little
	100	Stain penetration	Most
		Opacity	Opaque
		Color / oxidation	Whitish
		Crystal	Transparent, small (⅛"), difficult to see

CR 331 Frit #3110 Base
(See Figures 76–79)

Frit #3110 FERRO	50	Temperature	C/9
Zinc oxide	25	Surface @ C/9	Gloss
Flint	18	Fluidity	Fluid
Titanium dioxide	7	Stain penetration	Most
	100	Opacity	Translucent when thin
			Opaque when thick
		Color / oxidation	Milky blue-white
		Crystal	White needle shape

CR 332 Sodium Silicate

Sodium silicate SS 56 (dry)	50	Temperature	C/9
Zinc oxide	25	Surface @ C/9	Gloss
Flint	18	Fluidity	Fluid
Whiting	7	Stain penetration	All
	100	Opacity	Transparent
		Color / oxidation	Clear
		Crystal	Snowflake, frost white

Colorant		Color of Glaze	Crystal
Nickel carbonate	4	Amber (17)	Blue (85)
Copper carbonate	3	Turquoise blue (84)	Transparent

CR 333 Titanium-Zinc Crystal

Sodium silicate SS 56 (dry)	50	Temperature	C/9–10
Flint	20	Surface @ C/9	Gloss
Zinc oxide	20	Fluidity	Some flow
Titanium dioxide	8	Stain penetration	All
Strontium carbonate	2	Opacity	Transparent
	100	Color / oxidation	Window clear
		Crystal	White needle

Colorant		Color of Glaze	Crystal
Nickel carbonate	3.0	Amber (140)	Green (limited number)
Cobalt carbonate	1.2	Blue and green areas	Blue
Uranium	6.0	Yellow	Light cream
Copper carbonate	3.0 ⎱ Grass green		Light green
Tin oxide	2.0 ⎰		
Molybdenum	6.0	Milky	Silver

CR 334 Steven McGovney, Spotty Blue

Kingman feldspar (old)	46.9	Temperature	C/10
Barium carbonate	19.7	Surface @ C/10	Semimatt
Kentucky old mine #4	9.4	Fluidity	Some
Whiting	8.3	Stain penetration	Very darks
Rutile	8.1	Opacity	Opaque
Zinc oxide	7.6		
	100.0		
Black iron oxide	4.0	Color / oxidation	Spotty rutile
		Crystal	Subtle crystal spotting

CR 335 Keystone Crystal

Keystone feldspar	46	Temperature	C/9
Zinc oxide	28	Surface @ C/9	Gloss
Whiting	13	Fluidity	Fluid
Flint	8	Stain penetration	None
Rutile	5	Opacity	Opaque
	100	Color / oxidation	Whitish
		Crystal	Whitish green

Colorant		Color of Glaze	Crystal
Cobalt carbonate	1.2	Broken light blue	Blue (83)
Copper carbonate (See Figures 79 and 80)	3.0	Sea foam greens	Gray-green
Manganese carbonate	3.0	Warm creams (24)	Light brown (17)
Iron oxide	2.0	Cool creams (8)	Medium brown (34)
Cobalt carbonate	1.2	Gray-greens	Light blue (85)
Chrome oxide	2.0		
Copper carbonate	3.0	Tanish (24)	Light brown (36)
Manganese carbonate	3.0		

CR 336 Over-glaze Crystal
(Apply over a gloss-glazed surface)

Flint	45	Temperature	C/9
Zinc oxide	25	Surface @ C/9	Gloss (2-hour soak)
Sodium carbonate	12	Fluidity	Little
Calcium carbonate	9	Stain penetration	All
Kaolin	5	Opacity	Translucent
Titanium dioxide	4	Color / oxidation	Slightly whitish
	100	Crystal	Small transparent

CR 337 Steven McGovney, Royce Blue Spot

Kingman feldspar	44.8	Temperature	C/9–10
Barium carbonate	18.7	Surface @ C/9	Soft matt
Kentucky old mine #4	8.9	Fluidity	Little
Whiting	7.9	Stain penetration	Very darks
Rutile	7.7	Opacity	Opaque
Zinc oxide	7.2		
Bone ash	4.8		
	100.0		
Copper carbonate	2.0	Color / oxidation	Royce blue spot
Cobalt oxide	1.0	Crystal	Spotting
Copper carbonate	0.7	Color / oxidation	Light Royce blue spot
Cobalt oxide	0.5	Crystal	Spotting

CR 338 Broken Crystal Patterns

Keystone feldspar	44	Temperature	C/8–9
Zinc oxide	25	Surface @ C/8	Gloss
Whiting	12	Fluidity	Some
Flint	10	Stain penetration	Most
Yellow ochre	9	Opacity	Translucent
	100	Color / oxidation	Tans with some blues
		Crystal	⅛" crystals, bluish

CR 339 Soda Bicarbonate Base

Sodium bicarbonate	40	Temperature	C/9
Zinc oxide	30	Surface @ C/9	High gloss (2-hour soak)
Silica	30		
	100	Fluidity	Fluid
		Stain penetration	All
		Opacity	Transparent
		Color / oxidation	Water clear
		Crystal	White, ⅜"

Colorant		Color of Glaze	Crystal
Zircopax	5.0	Clear	White
Rutile	5.0	Cream (24)	White
Nickel carbonate	4.0	Amber (27)	Blue (95)
(See Figures 82 and 83)			
Copper carbonate	3.0	Light sea green	White
Cobalt carbonate	1.2	Medium blue (92)	Blue (92)
Manganese carbonate	6.0	Grayed pink	Clear
Black uranium oxide	6.0	Bright yellow (3)	Clear
Red iron oxide	2.0	Off-white	Clear
Ceramic black	8.0	Brown (137)	Clear
Nickel carbonate	2.0 ⎱	Light tan	Blue (85)
Copper carbonate	0.6 ⎰		
Cobalt carbonate	1.2 ⎱	Blue (89)	Clear
Manganese carbonate	6.0 ⎰		

CR 340 Oxford Feldspar Base

Oxford feldspar	40	Temperature	C/9
Zinc oxide	25	Surface @ C/9	Gloss (2-hour soak)
Wollastonite	20	Fluidity	Fluid
Soda ash	10	Stain penetration	Most
Titanium oxide	5	Opacity	Transparent
	100	Color / oxidation	Milky white
		Crystal	Snowflake, ⅜″

CR 341 Broken Gloss/Matt

Silica	38.8	Temperature	C/8–10
Zinc oxide	19.7	Surface @ C/10	Broken gloss/matt
Florida kaolin	16.7	Fluidity	None
Soda ash	10.7	Stain penetration	Darks
Whiting	6.1	Opacity	Opaque
Rutile	4.8	Color / oxidation	Tan and creams
Copper carbonate	3.2	Crystal	Very small, bluish
	100.0		

CR 342 John's Special
(See Figure 83)

Pearl ash	36.2	Temperature	C/8–10
Manganese dioxide	34.2	Surface @ C/9	Semigloss
Silica	29.6	Fluidity	Very fluid
	100.0	Stain penetration	None
		Opacity	Opaque
		Color / oxidation	Broken medium and dark browns
		Crystal	Medium crystals, browns

CR 343 Cream Crystals

Silica	35	Temperature	C/8–10
Zinc oxide	25	Surface @ C/9	Gloss
Soda ash	20	Fluidity	Very fluid
Whiting	10	Stain penetration	Some
Rutile	6	Opacity	Translucent and opaque
China clay	4	Color / oxidation	Whitish to light
	100		blues
		Crystal	Cream (5) crystals, opaque

CR 344 W. Pukall, Rutile Base Crystal

Red lead	34.9	Temperature	C/7
Silica	25.7	Surface @ C/7	Gloss/matt
Kaolin	15.8	Fluidity	Little
Rutile	10.1	Stain penetration	None
Zinc oxide	7.4	Opacity	Opaque/semitranslucent
Whiting	6.1	Color / oxidation	Grayish-violet
	100.0	Crystal	Compact needles, occasional
			star-shaped crystals
		Note: Must be fritted	

CR 345 W. Pukall, Star-Shaped Crystals

Silica	32.1	Temperature	C/7
Red lead	29.1	Surface @ C/7	Bright
Borax	14.6	Fluidity	Some
Feldspar	14.2	Stain penetration	Very darks
Barium carbonate	5.0	Opacity	Opaque
Potassium nitrate	2.6		
Whiting	2.5		
	100.1	Note: Must be fritted	

Rutile	15.0 ⎫	Color / oxidation	Olive
Copper oxide	6.0 ⎬	Crystals	Numerous fine greenish-
Manganese dioxide	3.0 ⎭		brown star-shaped crystals

CR 346 Interesting Crystal Pattern

Silica	32	Temperature	C/9–10
Nepheline syenite	28	Surface @ C/9	Semigloss/gloss
Zinc oxide	20	Fluidity	Some
Titanium dioxide	8	Stain penetration	Very darks
Strontium carbonate	8	Opacity	Opaque
Kaolin	4		
	100		

Copper carbonate	3.0 ⎫	Color / oxidation	Brown, light brown
Tungstic acid	1.0 ⎭		(36)
		Crystal	Medium green-brown all-over pattern

LOW-TEMPERATURE CRYSTAL GLAZES

Crystal glazes can produce crystals at either high or low temperatures, depending upon the glaze composition, maturing temperature, and cooling cycle. The crystals produced at low temperatures are well-defined individual crystals that usually cover the entire surface, touching each other much like galvanized-iron patterns (Figure 71). Lower-temperature glazes have high amounts of soda, boron, lead, or lithium fluxing agents. High-temperature glazes use fewer of these minerals and also contain potassium, calcium, zinc, and soda. The cooling cycle for low-temperature glazes is short, and a long soaking period is not necessary.

The following techniques apply to low-temperature crystal glazes.

Firing: The firing and cooling cycles for low-temperature crystal glazes are no different from those for any selected firing temperature. A thin-walled or fast-cooling kiln must have the first 300°F of cooling carefully controlled. It should take about one hour or more for the kiln temperature to drop 300°F; if the kiln begins to cool faster, adjustments must be made to slow the cooling process, either by turning the electric elements on to a low setting; or, in gas-fired kilns, the pilot burner or a low-flame setting is used. After the controlled cooling period, the rest of the cooling cycle is not monitored, and the kiln is allowed to cool at its own rate.

Clay body: High-iron or dark-firing clays are not recommended for use with crystal glazes because the clay materials will interfere with crystal production. Change of color, dry matt surface, or lack of crystal growth will result. White, light gray, or buff clay bodies give the best results, as they will not mute the color brilliance.

Glaze application: Raw borax, soda ash, sodium bicarbonate, boric acid, and other ingredients are granular and water-soluble. This makes glaze application by painting, spraying, and dipping difficult, because the soluble materials are absorbed into the clay body and build up an excessively thick glaze coating. Grinding the glaze by hand or with a ball mill, and adding suspension agents, can reduce the problem. Only enough glaze for one session should be prepared, for the glaze is hydraulic and will form a hard cake at the bottom of the glaze bucket in a few hours.

Cultivation of crystals on the surface of a fired glaze: Another method of obtaining crystals is to place certain mineral salts and sulphates directly on the surface of a high-fired glossy glaze. Cobalt sulphate, copper sulphate, chromium salt, magnesium sulphate, or other mineral salts are mixed with water and sprayed, dipped, or painted onto the glaze surface. When the solution dries, the salts will crystallize on the surface. The crystal pattern is then made permanent by a refiring of the glaze at 1000°F to 1600°F.

LOW-TEMPERATURE CRYSTAL GLAZE FORMULAS

The formulas are for C/07 and C/011, matt surface, generally opaque, little flow, and the crystals are a galvanized-iron type. The lead-based glazes are toxic and should not be used for making eating or drinking utensils.

CR 347 Red-Orange TOXIC
(See Figure 71)

Red lead	80	Temperature	C/07
Uranium oxide	15	Surface @ C/07	Matt
Barium carbonate	5	Fluidity	Little
	100	Stain penetration	Very darks
		Opacity	Translucent where thin
		Color / oxidation	Red-orange (67)
		Crystal	Large, clear glittering crystals

CR 348 Orange-Red TOXIC

Red lead	80	Temperature	C/07
Kaolin	14	Surface @ C/07	Matt
Uranium oxide	6	Fluidity	Some
	100	Stain penetration	Darks
		Opacity	Translucent where thin
		Color / oxidation	Orange-red, black where thick
		Crystal	Galvanized-iron effect

CR 349 Burnt Red

Borax	57.0	Temperature	C/011
Flint	28.6	Surface @ C/011	Matt
Red iron	14.4	Fluidity	None
	100.0	Stain penetration	None
		Opacity	Opaque
		Color / oxidation	Burnt red (75)
		Crystal	Galvanized-iron effect

Crystalline Matt Glazes

The satin surface of the most soft and beautiful matt glazes is the result of the interlacing of minute, needle-like crystals, whose major axes are approximately parallel to the surface of the glaze. The glaze, a nonshiny surface, is known as *butter matt, vellum matt, satin matt, velvet matt, zinc matt, smooth matt* and *crystalline matt.* When a matt glaze is carefully examined under a microscope, it will reveal a crystal structure belonging to one or more of the crystal groups or systems. The glaze is composed of exceedingly fine crystals that are not detectable with the eye or even at 10-power magnification. The crystals are so small that each structure is only a few molecules in size. If they are permitted to develop in a crystal-soaking period, however, they can become the nuclei for the growth of visible crystals. The chemical composition of the glaze at peak firing temperature is complex, and it changes upon cooling, when the matting agents crystallize into their more simple silicate forms. These silicates vary in composition, containing, for example, such minerals as zinc, calcium-zinc, or titanium-zinc. Matt finishes can also be obtained by the addition of dolomite, magnesium, talc, feldspar, bone ash, and kaolin, which are not absorbed completely into the "glaze melt" at the fired temperature. Underfiring a glaze or applying too thin a glaze coating will give the pot a *rough* matt-like surface. Although some of these finishes are smooth, they are not as attractive as the crystalline matts, lacking their soft appearance and tactile qualities.

COMPOSITION

To the base glaze one or more items—opacifiers, colorants, matting agents—can be added. Most gloss and slightly flowing base glazes can be converted into crystalline matt glazes by the addition of a matting agent. A *base glaze* is analogous to the house painter's white paint that serves as a base to which colorants are added to obtain the desired shade. The base glaze—which is colorless—is often translucent or transparent. *Opacifiers* (tin, zircopax, opax, zinc, and titanium, for example) can be added to lend various degrees of opacity to the base glaze; and *metallic oxides* (cobalt, nickel, iron, chrome, or other oxides) or *stains* can be added to provide color (see Chapter III). Matting agents can also be added, their amount varying with the nature of the base glaze and its firing temperature.

Matting Agent	Percent Range
Zinc oxide	5–30
Titanium dioxide	2–20
Rutile	4–20
Vanadium	1–10
Molybdenum	3–10
Magnesium	4–15
Barium carbonate	4–15
Ceramic black*	8–20
(or metallic oxides)	

* Mixture of cobalt, iron, copper, chrome, and manganese oxides.

Titanium, vanadium, or molybdenum are seldom used alone; rather, the matt formula will usually call for one-third of one of these agents and the two-thirds balance of zinc oxide. Barium matt is composed of a barium-based glaze (one that has more than 15 percent barium in its formula) plus a small amount of zinc to keep the surface smooth. Dolomite matt contains more than 5 percent magnesium in its base formula as its matting agent and contains a small amount of zinc or titanium to ensure a smooth matt surface. Rutile and titanium matts will often break up the color, creating streaked, marbled, or patterned glazes (Figure 72). Saturated metallic oxides like cobalt, iron, copper, chrome, or manganese—used by themselves or in a mixture of several oxides (which is the more usual method)—will produce a matt finish whose most notable characteristic is its smooth, gun-metal quality.

FIRING

The application and firing techniques used for crystalline matt glazes are not significantly different from those used for ordinary glazes. Because some of the matt glazes do not flow, an even application of glaze is essential if the application marks are not to show. Another significant consideration in matt glazing is the rate of cooling. The first 300°F of the cooling cycle must be controlled so that the decrease in temperature occurs at a slower-than-usual rate. This is not difficult in thick-walled kilns that naturally cool more slowly. Once the kiln has reached the maturation temperature, it is turned off, the dampers are closed, and the kiln is allowed to cool at its own rate. Allowing one to two hours for the temperature to drop the first 300°F is sufficient. In a gas kiln it may be necessary to keep the pilots on or the burner on at its lowest setting.* Whether this will be necessary depends upon the glaze used and kiln's cooling rate. After the kiln temperature has dropped by 400°F, the kiln can be allowed to cool at its own rate for the duration of the cooling cycle.

CRYSTALLINE MATT GLAZE FORMULAS

Various types of matt glazes are included in the following formulas. The temperature range of the formulas includes C/08, C/07, C/06, C/02, C/3, C/4, C/5, C/7, C/8, C/9, and C/10. Opacifiers and colorants can be added to these glazes if desired. The lead-based glazes are toxic and should not be used for making drinking or eating utensils.

CR 350 Broken Orange/Reds TOXIC

Red lead	83	Temperature	C/07
Uranium oxide	10	Surface @ C/07	Matt
Molybdenum	7	Fluidity	Little
	100	Stain penetration	Darks
		Opacity	Translucent where thin
			Opaque where thick
		Color / oxidation	Broken orange to reds
		Note: Grind, apply evenly in medium thickness.	

* An electric kiln should have one of the electrical controls turned on to the lowest setting.

CR 351 Frit #2106 Matt

Frit #2106 HOMMEL	65	Temperature	C/10
Zinc oxide	25	Surface @ C/10	Smooth semimatt
Silica	5	Fluidity	Little
Strontium	5	Stain penetration	All
	100	Opacity	Translucent
		Color / oxidation	Whitish (7)

CR 352 Buckingham Matt

Buckingham feldspar	63.8	Temperature	C/9–10
Zinc oxide	23.1	Surface @ C/9	600-mesh everycloth
Whiting	6.9		matt
Flint	6.2	Fluidity	None
	100.0	Stain penetration	All
		Opacity	Translucent
		Color / oxidation	White (7)
		Note: Apply evenly.	

CR 353 Rutile Brown Crystalline Matt TOXIC
(See Figure 72)

Red lead	63.0	Temperature	C/08–07
Soda ash	10.0	Surface @ C/07	Patchy gloss/matt
Flint	8.2	Fluidity	Fluid
Rutile	6.8	Stain penetration	Very darks
Soda feldspar	6.4	Opacity	Opaque
Red iron oxide	4.0	Color / gloss	Broken mustard yellow
Kaolin	1.6	/ matt	Brown
	100.0		

CR 354 Butter-Smooth Semimatt

Frit #399	60	Temperature	C/7–9
Zinc	40	Surface @ C/9	Butter-smooth
	100		semimatt (3-hour soak)
		Fluidity	Fluid
		Stain penetration	Darks
		Opacity	Opaque
		Color / oxidation	Subtle white (7)

CR 355 Dolomite-Zinc Matt

Buckingham feldspar	60	Temperature	C/9
Dolomite	20	Surface @ C/9	Matt (3-hour soak)
Zinc oxide	18	Fluidity	Little
Kaolin	2	Stain penetration	Most
	100	Opacity	Translucent
		Color / oxidation	White
Cobalt carbonate	1.0	Color / oxidation	Clean blue (95)
Chrome oxide	1.5	Color / oxidation	Mauve (darker 47)
Nickel carbonate	4.0	Color / oxidation	Turquoise (85)
Manganese carbonate	3.0	Color / oxidation	Pinkish (47)

CR 356 Frost Matt

Kona F-4 feldspar	60	Temperature	C/9–10
Barium carbonate	20	Surface @ C/9	Matt
Soda ash	8	Fluidity	Little
Flint	6	Stain penetration	All
Zinc oxide	6	Opacity	Transparent
	100	Color / oxidation	Clear with a soft frost-glittering matt

CR 357 Oxford Matt

Oxford feldspar	55	Temperature	C/9–10
Zinc oxide	18	Surface @ C/9	Semimatt
Silica	18	Fluidity	Some flow
Whiting	9	Stain penetration	Most
	100	Opacity	Translucent
		Color / oxidation	White

CR 358 Barium Matt

Nepheline syenite	52	Temperature	C/5–6
Barium carbonate	22	Surface @ C/5	Matt
Whiting	19	Fluidity	None
Molybdenum oxide	3	Stain penetration	All
Rutile	2	Opacity	Translucent
Lithium carbonate	2	Color / oxidation	White with surface sparkles
	100		

CR 359 Smooth Matt

Frit #3110 FERRO	50	Temperature	C/7–9
Zinc oxide	38	Surface @ C/9	Smooth matt
Flint	12	Fluidity	Fluid
	100	Stain penetration	All
		Opacity	Translucent
		Color / oxidation	White where thick

CR 360 Bright White Matt

Kona F-4 feldspar	50	Temperature	C/9–10
Silica	20	Surface @ C/9	Smooth matt
Zinc oxide	20	Fluidity	None
Pearl ash	10	Stain penetration	Very darks
	100	Opacity	Opaque
		Color / oxidation	Bright white
		Note: Apply even coverage.	

CR 361 Matt

Cornwall stone	48	Temperature	C/9–10
Flint	24	Surface @ C/9	Smooth matt
Zinc oxide	24	Fluidity	None
Dolomite	4	Stain penetration	Very darks
	100	Opacity	Opaque
		Color / oxidation	Whitish
		Note: Apply even coverage.	

CR 362 Satin Matt

Keystone feldspar	47	Temperature	C/8–10
Zinc oxide	34	Surface @ C/9	Satin matt
Whiting	12	Fluidity	Some
Flint	7	Stain penetration	Most
	100	Opacity	Translucent
		Color / oxidation	Warm whitish (64)

CR 363 29A

Oxford feldspar	46.2	Temperature	C/8–10
Zinc oxide	28.1	Surface @ C/9	Semimatt
Flint	13.2	Fluidity	Some
Whiting	9.0	Stain penetration	Darks show
Rutile	3.5	Opacity	Opaque
	100.0	Color / oxidation	White (7)

CR 364 Silica Matt

Silica	45	Temperature	C/06–04
Lithium carbonate	35	Surface @ C/06	Smooth matt
Kaolin	10	Fluidity	None
Soda ash	10	Stain penetration	Most
	100	Opacity	Translucent
		Color / oxidation	Slight white
Copper carbonate	3.0	Color / oxidation	Turquoise
Cobalt carbonate	0.5	Color / oxidation	Light blue
Red iron oxide	5.0	Color / oxidation	Yellowish

CR 365 Tortilla (La Jolla Museum of Art)

Kingman feldspar (old		Temperature	C/10
Kingman)	38.8	Surface @ C/10	Smooth matt
Cornwall stone	19.4	Fluidity	None
Whiting	19.4	Stain penetration	All
Kentucky old mine #4	19.4	Opacity	Translucent
Zinc oxide	3.0	Color / oxidation	Slight white
	100.0	Note: Some cracks where thick	

CR 366 Velvet Matt

Silica	38	Temperature	C/8–10
Zinc oxide	30	Surface @ C/9	Velvet matt
Soda ash	20	Fluidity	Some
Whiting	10	Stain penetration	All
China clay	2	Opacity	Translucent
	100	Color / oxidation	Milky

CR 367 Smooth Satin Matt

Kingman feldspar (old		Temperature	C/02
Kingman)	38	Surface @ C/02	Smooth satin matt
Barium carbonate	30	Fluidity	None
Flint	16	Stain penetration	Most
Whiting	8	Opacity	Translucent
Zinc oxide	6	Color / oxidation	Whitish
Lithium carbonate	2		
	100		

CR 368 Butter Matt

Custer feldspar	36.0	Temperature	C/8–10
Flint	26.0	Surface @ C/10	Butter matt
Magnesium carbonate	15.0	Fluidity	None
Whiting	13.5	Stain penetration	Darks
Tennessee ball	5.0	Opacity	Opaque
Zinc oxide	2.5	Color / oxidation	White
Barium carbonate	2.0		
	100.0		

CR 369 Zinc Matt

Sodium bicarbonate	35	Temperature	C/9
Zinc oxide	35	Surface @ C/9	Semimatt
Silica	30	Fluidity	Some
	100	Stain penetration	Most
		Opacity	Translucent
		Color / oxidation	White (7)

CR 370 Velvet Matt

Silica	34	Temperature	C/4
Sodium bicarbonate	30	Surface @ C/4	Velvet matt
Zinc oxide	21	Fluidity	Fluid
Boric acid	14	Stain penetration	All
Cobalt carbonate	1	Opacity	Translucent
	100	Color / oxidation	Medium cobalt blue
			with sugar-like sparkles

CR 371 Soft Matt

Whiting	32	Temperature	C/3–6
Kingman feldspar (old		Surface @ C/4	Soft matt
Kingman)	30	Fluidity	Fluid
Zinc oxide	25	Stain penetration	None
Kaolin	9	Opacity	Opaque
Flint	4	Color / oxidation	White
	100		

Toxicology and
Safety in Ceramics

The number and variety of potential hazards to which individuals may be exposed in the ceramic studio are considerable. Certain ceramic materials, in fact, are so dangerous to the potter that they can cause death if they are not handled properly. Extensive toxicological research of the materials used in ceramics has made great strides in alerting and educating the public, industry, and ceramists to the dangers inherent in ceramic production. Through the cooperation of potters, of industry, of medical societies, and of government, precautionary measures have been developed for the safe use of ceramic materials and equipment. The safeguards listed here, which incorporate those measures, should be conscientiously followed at all times:

1. Prominent placement of posters warning of exposure to hazardous conditions and the labeling of all materials and equipment as to their possible dangers.
2. Availability of appropriate instructions for the correct use of equipment and the safe use of potentially dangerous substances.
3. Distribution and posting of pamphlets that list the early signs of poisoning (such as hypersensitivity, nausea, and

dizziness) and that outline the series of steps to be taken in the event that safety precautions fail or were not followed and poisoning has already occurred.

4. The ventilation of the grinding wheel, spray booth, and enclosed glaze and kiln rooms to prevent inhalation of dust, toxic substances, and/or fumes.

5. Careful storage and disposal of potentially hazardous materials and the safekeeping, under lock and key, of potentially dangerous equipment (pugmill, claymixer, etc.) so that they are not readily accessible to inexperienced persons, children, and/or visitors.

6. The grounding of all equipment to eliminate the danger of electric shock.

7. Placement of adequate guards and safety devices on all grinders, kilns, and power equipment of any type.

Toxic Materials

The following ceramic materials vary in degree of toxicity and toxic effect. The most common toxic effect is physical damage to the respiratory tract. Lead, asbestos, arsenic, and barium are the most widely recognized toxic materials and are probably the most dangerous.

Alumina dust can cause irritation to the respiratory tract when inhaled. Prolonged and/or repeated exposure can lead to fibrosis, emphysema, and pneumothorax, which are all forms of aluminosis (a lung disease similar to silicosis—see "Silica" entry).

Antimony compounds will cause dermatitis, conjunctivitis, or nasal-septum ulceration through direct contact, or by inhalation of dust and/or fumes.

Arsenic and its salts are highly toxic. Chronic poisoning can result in the degeneration of the liver and kidneys, pigmentation of the skin, herpes, and irritation of the gastrointestinal tract, which can progress to severe shock and ultimately to death. (Arsenic trioxide, used to make green enamels, is also used for rodent poison, insecticide, and weed killers.)

Asbestos fibers can cause skin irritation and, if inhaled, can result in pulmonary fibrosis (asbestosis), emphysema, or cancer.

Barium compounds are toxic if ingested or inhaled, and excessive amounts will cause violent diarrhea, convulsive tremors, and muscular paralysis.

Bismuth subnitrate is used as a luster; inhalation of its fumes will cause acute headaches.

Borax and *boric acid* are toxic if absorbed, ingested, or inhaled and may cause nausea, vomiting, diarrhea, convulsions, and a comatose state.

Cadmium compounds, if ingested, cause increased salivation, choking, vomiting, and diarrhea. Inhalation will cause coughing, headaches, vomiting, pneumonitis, and extreme restlessness and irritability.

Carbon monoxide results from the incomplete combustion of firing fuel. In a closed room heavy fumes will concentrate at low levels. Mild exposure causes headache, dizziness, weakness, and nausea. Extensive exposure will cause collapse, unconsciousness, and death through lack of oxygen.

Chlorine gas is one of the exhaust gases that emerge from the stack during the salting phase of salt glazing. Firing in a closed kiln room or during windless, low-pressure weather will permit the gas to remain in the kiln area. Such a concentration of chlorine gas is irritating to the skin and to the mucous membranes of the respiratory tract. Exposure to a heavy concentration of the gas will cause fatal pulmonary edema.

Cobalt dust exposure may cause pulmonary distress symptoms and dermatitis.

Feldspar dust exposure is hazardous, because the dust contains free silica, whose absorption weakens the body's natural defense mechanisms.

Fiberglas fibers, like asbestos fibers, can cause skin irritation through direct exposure and, if inhaled, can result in pulmonary fibrosis or emphysema.

Iron chromate dust inhalation can lead to pneumonia.

Kaolin (china, ball, fire, and other clays, as well as Albany slip and others) contains free silica, and the symptoms of kaolin poisoning are the same as those of silicosis (see "Silica" entry).

Lead compounds are nearly all poisonous, with the exception of most frits. Inhalation is the most harmful mode of absorption. Lead is an accumulative poison. Small amounts of lead absorbed into the body can cause anorexia, vomiting, and convulsions. Continued lead absorption can produce permanent brain damage and even death.

Lithium compounds seldom produce fatal results from exposure. If they are ingested, however, they may cause kidney damage.

Manganese poisoning results from inhalation of manganese dust or fumes. Sleepiness, weakness, and paralysis may occur.

Mica compounds, dust in particular, are irritants to the respiratory tract.

Nickel may cause dermatitis in sensitive individuals through direct contact.

Selenium exposure may cause depression or dermatitis. Prolonged exposure to selenium dust, or its ingestion, will cause acute poisoning.

Silica (flint, quartz) by itself or as a free agent in feldspar and other clay compounds, can cause silicosis if inhaled. Silicosis is a disease of the lungs in which a chronic shortness of breath is caused by fibrosis of the lungs.

Stannic chloride used for fuming in the kiln (to produce the mother-of-pearl effect) is an irritating vapor that may injure the eyes and the mucous membranes of the respiratory tract.

Talc dust inhaled for a prolonged period can cause pulmonary fibrosis.

Uranium salts and compounds are highly toxic. Dermatitis, renal damage, carcinoma, and eventually death result from prolonged ingestion and/or inhalation.

Vanadium pentoxide is a respiratory irritant when inhaled, and can cause dermatitis through prolonged contact with the skin.

Zinc oxide in the form of dust and/or fumes may, if inhaled, cause metal fume fever.

Since any dust, regardless of its composition, can be a source of irritation to the respiratory tract, preventive dust-control measures should be taken. Such measures should include the immediate wiping up of spills; the regular cleaning of floors and work areas; the wearing of dust respirators when mixing dry materials; the stirring of glazes with wire whisks or ladles instead of with one's hands; and the installation of exhaust systems for spray booth and enclosed glaze- and clay-making areas. "An ounce of prevention is worth a pound of cure" is indeed sage advice in the ceramist's studio.

Temperature Equivalents for Orton Standard Pyrometric Cones
As Determined at the National Bureau of Standards

Large Cones

Cone Number	°C	°F	°C	°F
022	585	1085	600	1112
021	602	1116	614	1137
020	625	1157	635	1175
019	668	1234	683	1261
018	696	1285	717	1323
017	727	1341	747	1377
016	764	1407	792	1458
015	790	1454	804	1479
014	834	1533	838	1540
013	869	1596	852	1566
012	866	1591	884	1623
011	886	1627	894	1641
010	887	1629	894	1641
09	915	1679	923	1693
08	945	1733	955	1751
07	973	1783	984	1803
06	991	1816	999	1830

Small Cones

Cone Number	°C	°F
022	630	1165
021	643	1189
020	666	1231
019	723	1333
018	752	1386
017	784	1443
016	825	1517
015	843	1549
014	870	1596
013	880	1615
012	900	1650
011	915	1680
010	919	1686
09	955	1751
08	983	1801
07	1008	1846
06	1023	1873

05	1031	1888	1046	1915	1062	1944
04	1050	1922	1060	1940	1098	2008
03	1086	1987	1101	2014	1131	2068
02	1101	2014	1120	2048	1148	2098
01	1117	2043	1137	2079	1178	2152
1	1136	2077	1154	2109	1179	2154
2	1142	2088	1162	2124	1179	2154
3	1152	2106	1168	2134	1196	2185
4	1168	2134	1186	2167	1209	2208
5	1177	2151	1196	2185	1221	2230
6	1201	2194	1222	2232	1255	2291
7	1215	2219	1240	2264	1264	2307
8	1236	2257	1263	2305	1300	2372
9	1260	2300	1280	2336	1317	2403
10	1285	2345	1305	2381	1330	2426
11	1294	2361	1315	2399	1336	2437
12	1306	2383	1326	2419	1355	2471

Temperature Equivalents for Orton Standard Pyrometric Cones—Continued
As Determined at the National Bureau of Standards

Cone Number	Large Cones				Cone Number	P.C.E. Cones	
	1306°C	2383°F	1326°C	2419°F		150°c	270°F
12	1306°C	2383°F	1326°C	2419°F	12	1337°C	2439°F
13	1321	2410	1346	2455	13	1349	2460
14	1388	2530	1366	2491	14	1398	2548
15	1424	2595	1431	2608	15	1430	2606
16	1455	2651	1473	2683	16	1491	2716
17	1477	2691	1485	2705	17	1512	2754
18	1500	2732	1506	2743	18	1522	2772
19	1520	2768	1528	2782	19	1541	2806
20	1542	2808	1549	2820	20	1564	2847
23	1586	2887	1590	2894	23	1605	2921
26	1589	2892	1605	2921	26	1621	2950
27	1614	2937	1627	2961	27	1640	2984
28	1614	2937	1633	2971	28	1646	2995
29	1624	2955	1645	2993	29	1659	3018

Cone					Cone		
30	1636	2977	1654	3009	30	1665	3029
31	1661	3022	1679	3054	31	1683	3061
31½	—	—	—	—	31½	1699	3090
32	1706	3103	1717	3123	32	1717	3123
32½	1718	3124	1730	3146	32½	1724	3135
33	1732	3150	1741	3166	33	1743	3169
34	1757	3195	1759	3198	34	1763	3205
35	1784	3243	1784	3243	35	1785	3245
36	1798	3268	1796	3265	36	1804	3279
37	ND	ND	ND	ND	37	1820	3308
38	ND	ND	ND	ND	38	1850	3362
39	ND	ND	ND	ND	39	1865	3389
40	ND	ND	ND	ND	40	1885	3425
41	ND	ND	ND	ND	41	1970	3578
42	ND	ND	ND	ND	42	2015	3659

With permission of the Edward Orton Jr. Ceramic Foundation, 1445 Summit St., Columbus, Ohio 43201.

Bibliography of Suggested Readings

General Information

Andrews, A. I. *Ceramic Tests and Calculations.* New York: John Wiley and Sons, 1955.

Ball, F. Carlton. *Syllabus for Advanced Ceramics.* Bassett, Calif.: Keramos Books, 1972.

——————. *Syllabus for Beginning Pottery.* Bassett, Calif.: Keramos Books, 1971.

Behrens, Richard. *Ceramic Glazemaking.* Columbus, Ohio: Professional Publications, 1976.

——————. *Glaze Projects.* Columbus, Ohio: Professional Publications, 1972.

Binns, Charles. *The Potters's Craft.* New York: Van Nostrand, 1947.

Cardew, Michael. *Pioneer Pottery.* New York: St. Martin's, 1971.

Conrad, John. *Ceramic Formulas: The Complete Compendium.* New York: Macmillan, 1973.

——————. *Ceramic Manual.* Englewood Cliffs, N.J.: Prentice-Hall, 1980.

——————. *Contemporary Ceramic Techniques.* Englewood Cliffs, N.J.: Prentice-Hall, 1978.

Counts, Charles. *Pottery Workshop.* New York: Macmillan, 1973.

Dreisbach, Robert. *Handbook of Poisoning.* Los Altos, Calif.: Lange Medical Publishing, 1969.

Facts About Lead Glazes for Art Potters and Hobbyists. New York: Lead Industries Association, 1972. [Booklet]

Fairhall, Lawrence T. *Industrial Toxicology.* New York: Hafner, 1969.

FDA Laboratory Information Bulletin, No. 834. Food and Drug Administration, Division of Compliance Programs, Bureau of Foods. Washington: Government Printing Office.

Fournier, Robert. *Illustrated Dictionary of Practical Pottery.* New York: Van Nostrand Reinhold, 1977.

Fraser, Harry. *Glazes for the Craft Potter.* New York: Watson-Guptill, 1974.

Goldberg, Steven. *Glaze Calculation.* San Jose, Calif.: Billiken Press, 1972.

Grebanier, Joseph. *Chinese Stoneware Glazes.* New York: Watson-Guptill, 1975.

Green, David. *Understanding Pottery Glazes.* London: Faber and Faber, 1963.

Lawrence, W. G. *Ceramic Science for the Potter.* Radnor, Pa.: Chilton, 1972.

Leach, Bernard. *A Potter's Book.* London: Faber and Faber, 1960.

Moeschlin, Sven. *Poisoning, Diagnosis and Treatment.* New York: Grune and Stratton, 1965.

Nelson, Glenn. *Ceramics, A Potter's Handbook.* New York: Holt, Rinehart and Winston, 1978.

Noble, Joseph Veach. *The Technique of Painted Attic Pottery.* New York: Watson-Guptill, 1965.

Norton, F. C. *Ceramics for the Artist Potter.* Reading, Mass.: Addison-Wesley, 1956.

Parmelee, C. W., and Cameron G. Harman. *Ceramic Glazes.* Boston: Cahners, 1973.

Rhodes, Daniel. *Clay and Glazes for the Potter.* Radnor, Pa.: Chilton, 1973.

——————. *Stoneware and Porcelain: The Art of High Fired Pottery.* Radnor, Pa.: Chilton, 1959.

Sanders, Herbert H. *Glazes for Special Effects.* New York: Watson-Guptill, 1974.

Shafer, Thomas. *Pottery Decoration.* New York: Watson-Guptill, 1976.

Shaw, Kenneth. *Ceramic Colors and Pottery Decoration.* New York: Praeger, 1962.

Wood, Nigel. *Oriental Glazes.* London: Pitman, 1978.

Articles on Crystal Glazes

Brass, A. "Crystalline Glazes." *Transactions of American Ceramic Society,* No. 6 (1904), pp. 20–25.

Evans, Ronnie R. "Formation of Crystalline Glazes." Master of Arts thesis, California State University, San Diego (1973).

Haldeman, V. K. "Aventurine Glazes." *Journal of American Ceramic Society,* No. 7 (1924), pp. 824–33.

Hansen, Marc. "Zinc Silicate Crystal Glazes." *Ceramics Monthly,* December 1975, pp. 12–15.

Heubach, A. R. "Notes of Causes of Matness in Glazes." *Transactions of American Ceramic Society,* No. 15 (1913), p. 591.

Hughan, A. "Early Chinese Ceramic Glazes." *Ceramic Age,* No. 56 (1950), p. 214.

Koerner, J. "New Crystalline Glazes." *Transactions of American Ceramic Society,* No. 10 (1908), pp. 61–64.

Kondo, Seiji. "Colored Zinc Crystal Glazes." *Journal of Japanese Ceramic Association,* No. 33 (1925), pp. 188–202.

―――――――. "Crystal Glazes Based on Composition of Minerals: I." *Journal of Japanese Ceramic Association,* No. 33 (1925), pp. 399–418.

―――――――. "Manganese Crystal Glaze." *Journal of Japanese Ceramic Association,* No. 33 (1925), pp. 387–98.

Kramer, H. M. "Colors in a Zinc Silicate Glaze." *Journal of American Ceramic Society,* No. 7 (1924) pp. 868–77.

Lorah, J. R. "Uranium Oxide Colors and Crystals in Low Temperature Glaze Combination." *Journal of American Ceramic Society,* No. 10 (1927), pp. 813–20.

Mellor, J. W. "The Cultivation of Crystals on Glazes." *Transactions of British Ceramic Society,* No. 36 (1937), p. 13.

Norton, F. H. "The Control of Crystalline Glazes." *Journal of American Ceramic Society,* No. 20 (1937), pp. 217–24.

Parmelee, C. W., and J. S. Lathrop. "Aventurine Glazes." *Journal of American Ceramic Society,* No. 7 (1924), pp. 567–73.

Pearson, B. M. "Ceramic Glazes and Enamels." *Ceramic Age,* No. 55 (1950), pp. 45–46.

Pence, F. K. "Mat Glazes." *Transactions of American Ceramic Society,* No. 15 (1913), p. 413.

―――――――. "Theory for the Cause of Matness in Glazes." *Transactions of American Ceramic Society,* No. 14 (1912), p. 682.

Potts, A. P. "Notes on Mat Glazes." *Transactions of American Ceramic Society,* No. 13 (1913), p. 628.

Pukall, W. "My Experience with Crystal Glazes." *Transactions of American Ceramic Society,* No. 19 (1908), pp. 183–215.

Purdy, R. C. "Mat Glazes." *Transactions of the American Ceramic Society,* No. 14 (1912), p. 671.

―――――――, and Junius F. Krehbiel. "Crystalline Glazes." *Transactions of American Ceramic Society,* No. 9 (1907), pp. 319–407.

Rand, C. C., and H. G. Schurecht. "A Type of Crystalline Glaze at Cone 3." *Transactions of American Ceramic Society,* No. 16 (1914), pp. 342–63.

Ruddle, Frank H. "A Few Facts Concerning the So-called Zinc Silicate Crystals." *Transactions of American Ceramic Society,* No. 8 (1906), pp. 336–52.

Schurecht, H. G. "Experiments on Aventurine Glazes." *Journal of American Ceramic Society*, No. 3 (1920), pp. 971–77.

Snair, David. "Making and Firing Crystalline Glazes." *Ceramics Monthly*, December 1975, pp. 21–26.

Stull, Ray T. "Notes on the Production of Crystalline Glazes." *Transactions of American Ceramic Society*, No. 6 (1904), pp. 186–97.

Thiemecker, H. "Notes on Cone 10 Raw Crystal Glazes." *Journal of American Ceramic Society*, No. 17 (1934), pp. 359–62.

Worchester, Wolsey G. "The Function of Alumina in a Crystalline Glaze." *Transactions of American Ceramic Society*, No. 10 (1908), pp. 450–83.

Zimmer, W. H. "Crystalline Glazes." *Transactions of American Ceramic Society*, No. 6 (1904), pp. 20–25.